'A sense of catastrophe shapes the present. Terrorism and "war on terror", environmental collapse, pandemic and 'migrant crisis" build the background of the analysis pursued in this timely and original book. Investigating Western visual culture and imaginaries, Gaia Giuliani gives us a breathtaking tour across landscapes populated by monstrous creatures that, far from simply being the "West's" Other, continue to haunt it and in a way foreshadow the possibility of its vanishing. In the time of the catastrophe racialised bodies continue to be targeted by violent measures of control to allow the reproduction of the European and Western "we", as Giuliani effectively shows. But with a classical postcolonial move she is also able to grasp and expose the cracks and fissures that destabilize that "we" and open up the space for a postcolonial and feminist political project built upon such notions as "trans-corporeality", "interactivity" and "interdependency of Life and Nonlife". Working the boundary between postcolonial, visual, and film studies, and at the same time drawing upon a number of other fields of knowledge, including philosophy and political theory, *Monsters, Catastrophes, and the Anthropocene* is a masterful academic work and a powerful contribution to a critical theory of our present predicament.'

Sandro Mezzadra, Professor of Political Theory, University of Bologna, Italy, co-author (with Brett Neilson) of Border as Method, or, the Multiplication of Labor *(Duke, 2013)*

'In geological time scale, Anthropocene is a period from WWII to current time. In Gaia Giuliani's extraordinarily erudite book, it is a time defined by "ontologies and logics" of Othering, where the Western/European "we" produces fears through mediatized narratives of monstrosity and catastrophe as its existential threats. Those fears are supposed to keep at bay resistance to extractivism of Earth's resources and neoliberal exploitation and exclusion of people deemed expendable. Racialized, gendered, sexualized and classed constructions of monsters serve to preserve and continue colonial-cum-capitalist technologies of power and their political, social, economic and cultural outcomes that privilege white bodies while simultaneously inflicts cultural, material and mortal violence on all others. Giuliani's critical feminist, postcolonial and ecological perspective offers an exceptional intersectional and genealogical analysis of plentiful examples from political theory and cultural production that links representations of contemporary migration, terrorism and natural disasters to the old colonial tales and images of slavery, apocalypse and endless forms of dehumanizing violence. Importantly, Giuliani also offers a glimpse of political practice that would link human and non-human life with non-living environment in different, non-exploitative modes of production (de-growth, non-exploitative) as well as social reproduction marked by interdependency of self-care and earth-care.'

Dubravka Zarkov, retired, Associate Professor of Gender, Conflict, Development, ISS/Erasmus University Rotterdam, The Netherlands, Research Associate, Radboud University Nijmegen, co-editor, European Journal of Women's Studies, *https://journals.sagepub.com/home/ejw*

'*Monsters, Catastrophes and the Anthropocene: A Postcolonial Critique* is an urgent and rigorous theorization of global regimes of extractive capitalism, environmental devastation, pandemics, and ongoing war and state violence. Giuliani offers fresh and insightful ways of approaching crises and reimagining what belonging could be like if *we* abandon a notion of "we" that has promulgated exclusion, suffering, and the deaths of many millions.'

Nicole R. Fleetwood, Professor of American Studies and Art History, Rutgers University, USA, author of Marking Time: Art in the Age of Mass Incarceration *(Harvard UP, Spring 2020)*

'Gaia Giuliani's work has been consistently marked by an ambitious level of engagement and impressive scope. *Monsters, Catastrophes and the Anthropocene: A Postcolonial Critique* is no exception. It takes a long, hard view of cultural history with particular reference to Europe/the West/Global North relations with its others. Giuliani places monsters as the prism through which these engagements unfold across time. But also how monsters inhabit our own ghostly crises times, culminating in the book's conclusion on COVID-19. The result is a broad-scoped analysis of "cultural texts" with particular attention to popular culture.'

Lars Jensen, Associate Professor of Intercultural studies, Department of Communication and Arts, Roskilde University, Denmark, author of Postcolonial Europe *(Routledge, 2020)*

'Giuliani's extremely rich volume offers precious keys to decipher salient features of the current "end times", murky and agitated by the spectre of planetary crises and permanent catastrophe. Leading the reader through an intense and exciting voyage in recent works of fiction, Giuliani skilfully traces the mutations of classic tropes of environmental discourse (contagion, crises, catastrophe and collapse), and their proliferation in contemporary political debates. Giuliani casts a spotlight on liminal figures such as the alien, the mutant, the monster, all situated in the in-betweens dead/undead, human/non-human. Embodied in widespread representations of the migrant, the terrorist, the victim of climate change/disaster, such figures of the monstrous are analysed by Giuliani as symptom of the reconfiguration of the boundaries between Life and Nonlife, a key site of political contestation in the face of tangled planetary crises. With a thorough and theoretically engaged exploration of visual imaginaries, Giuliani shows how apocalyptic (environmental) narratives extend into the future the postcolonial, racialised, gendered and classed relations that structure current fears and visions. Putting in conversation political theory, environmental humanities, postcolonial and critical feminist studies, Giuliani's is a thought-provoking intervention in critical debates on the Anthropocene, and a contribution to the pursuit of non-exploitative, caring and decolonized constellations of (non)human Life/Nonlife.'

Giovanni Bettini, Lecturer in Climate Politics and Development, Lancaster Environment Centre, Lancaster University, UK

'With clear insight of the fringes and structural contradictions of our time, Giuliani's analysis celebrates an investigative method developed through her long-standing research. Through a broad comparative analysis of visual apocalyptic materials, such as films and TV series, to trace protagonists of present popular imaginary, Giuliani offers original reflections and a visionary energy toward a postcolonial critical approach to contemporary fears of the End. Through a careful use of diverse disciplinary registers, Giuliani's book innovates philosophical form by building a historical and symbolic journey through the space–time geography of the world, masterfully braiding the threads of the colonial past and neo-colonial present to fix its knots in the construction of figures *at the border* of social fear; from the monster to the alien, from the virus to environmental catastrophes. The inevitability of concluding on the occasion of the COVID-19 pandemic perfectly closes the circle of reflection, stigmatizing our time and future as an era of a realized (capitalistic) dystopia. The culmination of Giuliani's brilliant book, however, is nested in its luminous ability to incite ways to think and move toward "a feminist, post-developmental and ecologist epistemology and a political project that embraces a new conception of *the political*."'

Giovanni Ruocco, Associate Professor in History of Political Thought, Department of Political Science, University of Rome 'La Sapienza', Italy, author of Razze in teoria. La scienza politica di Gaetano Mosca nel discorso pubblico dell'Ottocento *(Quodlibet, 2017)*

Monsters, Catastrophes and the Anthropocene

Monsters, Catastrophes and the Anthropocene: A Postcolonial Critique explores European and Western imaginaries of natural disaster, mass migration and terrorism through a postcolonial inquiry into modern conceptions of monstrosity and catastrophe.

This book uses established icons of popular visual culture in sci-fi, doomsday and horror films and TV series, as well as in images reproduced by the news media to help trace the genealogy of modern fears to ontologies and logics of the Anthropocene. By logics of the Anthropocene, the book refers to a set of principles based on ontologies of exploitation, extermination and natural resource exhaustion processes determining who is worthy of benefiting from value extraction and being saved from the catastrophe and who is expendable. Fears for the loss of isolation from the unworthy and the expendable are investigated here as originating anxieties against migrants' invasions, terrorist attacks and planetary catastrophes, in a thread that weaves together re-emerging 'past nightmares' and future visions.

This book will be of great interest to students and academics of the Environmental Humanities, Human and Cultural Geography, Political Philosophy, Psychosocial Studies, Postcolonial Studies and Critical Race and Whiteness Studies, Gender Studies and Postcolonial Feminist Studies, Cultural Studies, Sociology, Cultural Anthropology, Cinema Studies and Visual Studies.

Gaia Giuliani is Researcher at Centro de Estudos Sociais, University of Coimbra, Portugal.

Routledge Environmental Humanities

Series editors:
Scott Slovic
University of Idaho, USA,
Joni Adamson
Arizona State University, USA and
Yuki Masami
Aoyama Gakuin University, Japan

The *Routledge Environmental Humanities* series is an original and inspiring venture recognising that today's world agricultural and water crises, ocean pollution and resource depletion, global warming from greenhouse gases, urban sprawl, overpopulation, food insecurity and environmental justice are all *crises of culture*.

The reality of understanding and finding adaptive solutions to our present and future environmental challenges has shifted the epicentre of environmental studies away from an exclusively scientific and technological framework to one that depends on the human-focused disciplines and ideas of the humanities and allied social sciences.

We thus welcome book proposals from all humanities and social sciences disciplines for an inclusive and interdisciplinary series. We favour manuscripts aimed at an international readership and written in a lively and accessible style. The readership comprises scholars and students from the humanities and social sciences and thoughtful readers concerned about the human dimensions of environmental change.

Monsters, Catastrophes and the Anthropocene

A Postcolonial Critique

Gaia Giuliani

Routledge
Taylor & Francis Group
LONDON AND NEW YORK

earthscan
from Routledge

First published 2021
by Routledge
2 Park Square, Milton Park, Abingdon, Oxon OX14 4RN

and by Routledge
605 Third Avenue, New York, NY 10017

First issued in paperback 2022

Routledge is an imprint of the Taylor & Francis Group, an informa business

British Library Cataloguing-in-Publication Data
A catalogue record for this book is available from the British Library

Library of Congress Cataloging-in-Publication Data
Names: Giuliani, Gaia, author.
Title: Monsters, catastrophes and the anthropocene : a postcolonial critique / Gaia Giuliani.
Description: Abingdon, Oxon ; New York, NY : Routledge, 2021. | Series: Routledge environmental humanities | Includes bibliographical references and index.
Identifiers: LCCN 2020023112 (print) | LCCN 2020023113 (ebook) | ISBN 9781138479777 (hardback) | ISBN 9781351064866 (ebook)
Subjects: LCSH: Fear–Social aspects. | Risk perception. | Disasters–Social aspects. | Emigration and immigration–Social aspects. | Terrorism–Social aspects. | Monsters in mass media. | Monsters–Symbolic aspects. | Horror films–Social aspects.
Classification: LCC HM1101 .G58 2021 (print) | LCC HM1101 (ebook) | DDC 302/.17–dc23
LC record available at https://lccn.loc.gov/2020023112
LC ebook record available at https://lccn.loc.gov/2020023113

ISBN: 978-0-367-61229-0 (pbk)
ISBN: 978-1-138-47977-7 (hbk)
ISBN: 978-1-351-06486-6 (ebk)

DOI: 10.4324/9781351064866

Typeset in Goudy
by Wearset Ltd, Boldon, Tyne and Wear

To all the monsters resisting the Anthropocene

Contents

Figures

Acknowledgements

Monsters Catastrophes and the Anthropocene: A Postcolonial Critique was a long time in gestation and the circumstances by which it came to be were especially challenging. Several events inspired the research on which the book is based, from the September 11 terror attacks to the so-called war on terror, from the establishment of border regimes in the Mediterranean and beyond to the rise of far-right parties and movements in Europe to the large-scale environmental catastrophes of the last 20 years. Some of my findings have already been published in books and articles. I am grateful to Fulvio Cammarano, Series Editor of Quaderni di Storia, and to Alessandro Mongatti for their support in the publication of the monographic book *Zombie, alieni e mutanti. Le paure dall'11 settembre ad oggi* (2016). Among my published articles are 'Il senso degli italiani (e degli europei) per l'Altrx che vien dal mare. Cinema, bianchezza e Middle Passage mediterraneo' with Francesco Vacchiano, in the special issue 'Narrazioni postcoloniali' edited by Leonardo de Franceschi and Farah Polato in *Imago* (2019); 'Monstrosity, abjection and Europe in the war on terror', in *Capitalism Nature Socialism* (2016), 'Fears of disaster and (post-) human raciologies in European popular culture (2001–2013)' in the special issue 'Cultures of Disasters' edited by Anders Ekström and Kyrre Kverndokk in *Culture Unbound* (2015). Several articles are forthcoming, including 'Online social media and the construction of sexual *moral panic* around migrants in Europe', with Sofia José Santos and Júlia Garraio, in *Socioscapes* (2020) and 'The end of the world as we know it' in the special issue 'Cidadania em perigo: crime, fim-do-mundo e biopolítica nas literaturas e no cinema pós-coloniais' I am editing for *ecadernos* (2020).

The book also draws on research that has been published as book chapters, including: 'Afterword: life adrift in a postcolonial world', in *Life Adrift: Climate Change, Migration, Critique* (2017), edited by Andrew Baldwin and Giovanni Bettini; 'The colour(s) of Lampedusa' (2017), in *Border Lampedusa. Subjectivity, Visibility and Memory in Stories of Sea and Land*, edited by Gabriele Proglio and Laura Odasso; 'The Mediterranean as a stage: borders, memories, bodies' (2016) in *Decolonising the Mediterranean: European Colonial Heritages in North Africa and the Middle East* edited by

Gabriele Proglio, and the forthcoming 'Monstrous beauties: bodies in motion between colonial archives and the migrant and refugee crisis' in the *Palgrave Handbook on Race and Gender*, edited by Shirley A. Tate. The first Section of the book draws on findings published as an article, 'Il mostro che viene dal sud del mondo', in *Jacobin Italia* 6 (2020). Findings from the Conclusions appeared on the Italian blog 'Studi sulla questione criminale', which in March 2020 launched a call for critical reflection on bio- and necropolitics in the time of the COVID-19 pandemic.

I would like to extend my gratitude to my family and friends for their emotional support and intellectual assistance in these frantic times: my parents, Giuliano and Patrizia, and my dearest friends in Italy, Portugal, Britain and Brazil. Their contributions, from suggestions to commentaries to films and opinions have been crucial for this book. They showed me love and consideration, especially when my project of unpacking the European and Western imaginary of the 'crises' of the present time seemed an impossible task. Many among them, along with Simona, Francesco and Stefano Maffeo, Fidelia Avanzato, Anna Pegna, Luca Onesti, Emerson Pessoa, Vincenza Perilli, Caterina Peroni, Elena Lolli, Chiara Martucci, Gea Piccardi, Carmine Cassino, Samuele Biagiotti, Vittoria, Daniela and Leo, Ambra Formenti, Paolo Gorgoni, Sabrina Marchetti, Cristina Morini, Caterina Peroni and Simona Ricci contributed to my ethnographic research, which is included in the conclusions to the book in the form of a chronicle from the 'we's' 'red zones', with pictures of deserted cities and queues outside supermarket, links to news articles and comments via phone, email, text messages and Facebook posts.

A special thanks goes to my friend and colleague Marco Armiero, who encouraged me to send a book proposal to Routledge, Environmental Humanities series, to my kind and helpful commissioning editor, Rosie Anderson, and to the senior editor, Fran Ford, for agreeing on the best possible title for this book. A special shoutout goes out to the Centre for Social Studies (CES) at the University of Coimbra, Portugal, where I am permanent researcher, and to Silvia Loffredo, translator and editor, who provided English language editing of a draft of this manuscript. And I would like to express my gratitude to the librarians at the CES library Norte/Sul for always going out of their way to get me the books I needed in the shortest time possible.

My most heartfelt thank goes to Giovanni Bettini, my first reviewer, who read the entire manuscript and offered invaluable comments and suggestions. I also benefited from the precious insights of many other colleagues who read parts of the manuscript and suggested essential readings or shared their work with me: Gianni Ruocco, Gianluca Buonaiuti, Silvia Rodeschini, Lars Jensen, Iain Chambers, Miriam Tola, Andrea Pogliano, Sofia José Santos, Stefania Barca, Emanuele Leonardi, Júlia Garraio, Tatiana Petrovich Njegosh, Luigi Lucci, Kiran Grewal, Francesca Esposito, Marilena Indelicato, Airleas Margherita Vaglio and Irina Velicu. Invaluable advice also came from Martina Tazzioli, Orazio Irrera, Luciano Nuzzo, Nicola Mai, Marco Armiero, Alessandra Ferrini and Nina Amelung.

Enduring support and much knowledge-sharing came from the team members of the research project I am leading at CES, '(De)Othering: Deconstructing Risk and Otherness: Hegemonic Scripts and Counter-Narratives on Migrants/Refugees and "Internal Others" in Portuguese and European Mediascapes' (2018–2121): Silvia Roque, Sofia José Santos, Júlia Garraio, Rita Santos and Marilena Indelicato, thank you. This book is one of the main outcomes of our project (Reference: POCI-01-0145-FEDER-029997), funded by the Portuguese Foundation for Science and Technology FCT (DL57/2016/CP1341/CT0025 and CES-SOC/UID/50012/2019) and by FEDER, the European Regional Development Fund, through the COMPETE 2020 Operational Programme for Competitiveness and Internationalization (POCI), and by other Portuguese institutions through the FCT.

I would like to thank everyone at the Inter-Thematic Research Group on Migration (ITM) at CES for their support, and namely my colleagues, students and friends Cristiano Gianolla, Sílvia Leiria Viegas, Joana Sousa Ribeiro, Carla Panico, Sílvia Roque, Fatíma Velez de Castro, Clara Keating, Marilena Indelicato and Gisele Almeida. The events in the 'Migrating Rights | keywords' series have provided me with much opportunity for reflection. I had the privilege of presenting my findings and having my work discussed by colleagues Clara Keating, Manuela Ribeiro Sanches and Júlia Garraio (19 February 2020) at two such events: 'Borders' with Carla Panico and Irina Velicu (11 December 2019), and 'Middle passages'. I would like to extend my gratitude to the 'Officina Ecologia e Sociedade' at CES – a constant source of inspiration and intellectual nourishment.

Over the past seven years, my colleagues Gianni Ruocco at Sapienza in Rome and Cristina Demaria at University of Bologna have given me the opportunity to lecture their undergraduate and master's students on selected topics from the book, allowing me to receive feedback from a vast, committed audience, and for this I am immensely grateful.

I am extremely grateful to the Birkbeck Institute for the Humanities at the University of London for awarding me a Fellowship in 2018 to organise three events on the book's topics – the workshop 'Monstrosity, Abjection and Climate Change in the Mediterranean Area' (25 May 2018), an open interdisciplinary panel on 'Lampedusa and Beyond' (4 June 2018) and the roundtable 'Monstrosity and the "Refugee/migrant Crises"' (12 June 2018). Among the guest speakers who joined me at Birkbeck were Giovanni Bettini, Calogero Giametta, Ben Gidley, Nadine El-Enany, Sarah Keenan, Roger Luckhurst, Lorenzo Pezzani, Martina Tazzioli, Sarah Turnbull and Agnes Woolley. A special thanks goes to Laura Posocco at Birkbeck, who supported my application for the Fellowship.

Earlier drafts of my research were presented over the last seven years at symposia and conferences held by the University of Oslo (2013), Chelsea

College, London (2014), the KTH Royal Institute of Technology in Stockholm (2016), Georg August Universität, Göttingen (2016), the University of Coimbra (2016 and 2018), Lisbon's Nova (2018) and the University of Reykjavik (2019).

A special thanks goes to Marcella Simoni and Davide Lombardo, who invited me to present my research on Italy and the Mediterranean border regime at the symposium 'Languages of Discrimination and Racism in Italy in the 20th Century' they organised in September 2018 at New York University, Florence campus; to the Galleria Nazionale d'Arte Moderna e Contemporanea and to IILA (the Italian-Latin American Institute) for inviting me to discuss my work on the monstrification of human mobility at the international feminist art Festival 'Women Out of Joint' in Rome; to my colleague Teresa Cunha and the organising committee at the 'Ecologias feministas de Saberes' Winter School, held at CES in January 2019, for their invitation to present an early version of my work on monstrosity and the colonial archive; to Nina Amelung, Sheila Khan and Helena Machado for asking me to present my research at the ERC EXCHANGE Seminars in February 2019; to Giulia Casalini and the organisers at CUNTemporary for inviting me to discuss my research at the symposium 'Ecofutures' at Queen Mary University of London in April 2019; and to Laura Favaretto and Chiara Figone for asking me to join the 'Amnesia' round table held at the 2019 Venice Biennale and part of artist Laura Favaretto's 'Clandestine Talks' series.

Many events I was expected to attend in the near future have now been cancelled because of the COVID-19 pandemic, including the April 2020 'Humanity/Humanities on the Move Conference' at the University of Minho in Braga, Portugal; the 'Feminists Care for the Political Conference' scheduled for May 2020 at Sabanci University Gender and Women's Studies Centre of Excellence in Istanbul; the conference 'Toward a History of Modern Colour' scheduled to take place at the end of May 2020 at the University of Cambridge; and the 'Border Criminologies' Seminar at the Faculty of Law, University of Oxford, where I was supposed to present my work in June 2020.

My heartfelt thanks and appreciation go to all the organisers, staff and colleagues.

Lisbon, 15 April 2020

Introduction

1 The Time we are living in

'We' are living in a time of monsters and catastrophes, as if propelled into a never-ending dystopian loop. The finis mundi *is getting closer and closer and gradually becoming the only lens through which Europe and the West give meaning to 'our' time. 'We' fear invasions, a permanent state of terror and the ultimate environmental catastrophe – 'our world' overflowing with chaos threatening the order that guarantees our safety, well-being, sustainability and progress. As in Saint John's apocalypse, the end of the world as 'we' know it will erase Time and Space and irremediably harm the human body, bringing back the unbounded violence that had been expelled from the space of reason. What is in danger is the very essence of* humans, *who are left unprotected and exposed to barbarity, epidemics and natural disasters against which borders, walls, colonies, segregated spaces, thicker identities and martial laws need putting in place: we will do all it takes to stop the spread of chaos and keep it 'outside', even if it means sacrificing some for the good of the many. Some are already paying the highest price, but it cannot be helped – their own lack of knowledge makes them vulnerable to disaster. If we manage to keep at a safe distance from toxic waste, viruses, environmental pollution, wars and other harmful effects of the same neoliberal capitalism we benefit from, the best of humanity will be safe.*

Unruly mobility from the Global South to the Global North, post-9/11 organised terrorism and the ever-evolving environmental crisis have unleashed a complex assemblage of anxieties, fears and apocalyptic discourses that are today as pervasive and, more or less, implicitly deployed in mainstream media narratives as in popular culture – and differentially reproduced by national and international actors involved in border control, counterterrorism and climate-change adaptation. The news media, for instance, have been instrumental in 'naming an enemy' in the coronavirus pandemic that began in China at the end of 2019, transforming anxiety into fears that fuelled worldwide hostility against the Chinese 'virus spreaders'. In a time of crisis, political parties and other organisations exploit widespread anxieties over an uncertain future for partisan advantage, and far-right movements have been particularly adept at using fear and apocalyptic narratives to draw in followers and resources (Giuliani et al. 2020a; Giuliani et al. 2020b).

When threats experienced by the 'we' are depicted as *global* and solutions are peddled as *universal*, they are more likely to be seen as coming from an objective, neutral source standpoint and as generalisable to humankind as a whole, despite the 'we' being the expression a specific positionality in the Global North. The impact of catastrophes, however, is not as colour-blind and gender-neutral as the 'we' would like to think, overlooking the fact that not only race and gender but also poverty, sexuality and nationality are key factors in determining who gets to be saved and who is left behind. Casualties are and will be inevitable, but the 'we' is not willing be held accountable for sacrificing the lives of those who are, in fact, the most vulnerable, the wretched of the Earth.

Concepts of chaos and order are a reflection of Western understandings of both *the political* and what is *beyond* politics. They interact with concepts of enmity and war, as well as with the figures of barbarian/infidel and monster that are foundational to the ancient notion of polis, medieval conceptions of the city of God, and early and late modern ideas of society, state and international relations. While the specific concepts may have changed across time and space, as Foucault (1999) argues for the monster, they have always been attached to an idea of *the political* as the arena in which the 'we' and 'they' engage but also negotiate conflict (Portinaro 1992): unlike barbarians and infidels within and outside the citadel, enemies are seen as *symmetric* entities (regular armies of mutually acknowledged states). Building on Carl Schmitt (1962), Reinhart Koselleck (1979) distinguished not only between internal and external conflicts but also between external 'regular' wars involving regular armies and external 'irregular' wars – such as colonial wars against barbarians. *Securitisation* and *normalisation* processes are implemented (Foucault 1976b) in *asymmetrical* internal conflicts against that which is criminalised as the internal enemy (e.g. partisan and insurgent warfare). Standards of international protection are disregarded in external irregular wars. In both cases, the *asymmetrical* conflict is based on the monstrification of the enemy, as there is a mutual recognition of status among *symmetric* entities from which the barbarian and the monster are excluded. Monsters are, by definition, expendable, because it is against them that the order is built and maintained lest they challenge social and political assumptions, eventually undermining the endurance of the body politic. In fact, they are:

> at the same time the effect and the bodily manifestation, and therefore the visible aspect of the crisis. In other words the monster reveals a character that is contingent and therefore arbitrary of social, political, and cultural distinctions through which identities are constituted. It puts them in doubt and interrogates them on their presumed naturalness.
>
> (Nuzzo 2013, p. 58)

Monsters as well as nature are differentially acknowledged within the space of *the political*: they are not subjects, let alone subjects of rights. As such, monsters in Western imagination seem to elude all classifications and rules. They are perceived as embodying their transgression, and their unruly mobility is troubling (Neocleous 2005, p. 28). For these reasons, they become the object of bio- and necropolitics. Nature, instead, is 'beyond politics': it is included in the body politic only inasmuch as it provides a habitat or the resources that support life.

Therefore, monsters and nature challenge the order in quite different ways: while monsters undermine the *naturalness* of arbitrary distinctions that sustain the 'imagined community' (*bios*, or social organisation), nature is 'the exact opposite of freedom' (or *zoe*, the biological structure of life) (Kant 1756, cited in Clark 2011, p. 90). Nature as *zoe*, if not mastered by humankind, can harm *bios* too. And according to the Cartesian dichotomy that opposes it to reason, nature delimits the boundaries of the subject's autonomy within *bios* (Esposito 2008; Braidotti 2013).

In Western modern political thought, these are generally conceived as *universal* concepts. Yet as Asad (2007) and Butler (2009) have argued in the context of the so-called global war on terror (Galli 2010), it is precisely this *universality* that is being questioned (Bhambra 2015): the September 11 attacks on the safe space of the 'we' clearly involved different dynamics from *symmetric* warfare, and the 'war on terror' that followed brought into international relations modes of warfare that had been typical of colonial wars. A new kind of enemy emerged, a *monstrified* personification of Otherness produced by the conflation of enemy and monster in the figure of the terrorist, to whom, as in colonial wars, no protection nor legal status was granted. What is more, the conflation of war and securitisation enabled the warfare apparatus to borrow from ontologies, technologies and techniques traditionally used against the internal monster: as Judith Butler reminds us, 'the humans who are imprisoned in Guantanamo do not count as human; they are not subjects protected by international law' (Butler 2004, pp. xv–xvi). In blowing up the distinctions between war and colonial war as well as between enemy and criminal, and erasing the spatial and symbolic distance between civilised and uncivilised, Here and Out there, the attacks revealed the limited and biased scope of that universality.

Today, the monster is feared to have returned in the guise of masses of migrants, refugees and terrorist cells striking back from both within and outside the *bios*. At the same time, nature is striking back, too, and making issues of climate change, ocean acidification, air pollution, toxic waste and devastated land invade the space of the 'we', threatening its *zoe*. The omen of planetary catastrophe reveals the impossible task of keeping the environment out of *the political*, since human supremacy over nature is no longer a given. The effects of both glocal wars and environmental catastrophes translate into masses of migrants heading towards the Here, trespassing the borders of the polis and blurring arbitrary 'social, political and cultural distinctions through which identities are constituted' (Nuzzo 2013, p. 58). Thousands of migrants and refugees have trespassed borders and walls 'protecting' the Here and the 'we' in the 20 years since 9/11, proving that their 'will to escape' – a source of profit for neoliberal surplus-value extraction – can be controlled and channelled but not stopped. *Monstrosity* gains a renewed signification in this troubled scenario, signalling a 'catastrophic time' (Stengers 2015) that is both *global* and *planetary*.

2 About this book

Monsters, Catastrophes and the Anthropocene: A Postcolonial Critique explores European and Western imaginaries of natural disaster, mass migration and terrorism through a postcolonial inquiry into modern conceptions of monstrosity and catastrophe. Established icons of popular visual culture in sci-fi, doomsday and horror films and TV series as well as in images reproduced by the news media help trace the genealogy of modern fears to ontologies and logics of the Anthropocene. Yet the book does not stop at unveiling the inherent violence of the Anthropocene but goes on to propose a feminist, post-developmental and ecologist epistemology and a political project that embraces a new conception of *the political*.

Figure I.1 Gaia Giuliani, 'Colonial lighthouse in Cape Verde', 3 March 2019, Island of São Vicente.

Ideas and constructions of monstrosity have been widely investigated in many fields, including art, literature, philosophy, history, cultural studies, and feminist and queer studies. My analysis conceives of monstrosity as a historically produced discursive process of *monstrification* that may reveal the relation between the operation of ontologies and logics of the Anthropocene and their legitimation across time and space.

My critique of monstrosity and catastrophe predominantly focuses on Europe and the West, whose imaginary of crisis – though differently articulated in scientific, political and popular culture within and across national boundaries – is shared by a transnational 'we' that, akin to an 'imagined community' (on my take on Benedict Anderson's concept [1983], see Giuliani 2019), is grounded in common experiences of *history*, *geography* and *humanness*. The 'we's' perspective and worldview instantiate the belief systems of white hegemonic culture based on a hierarchical reading of bodies, cultures, social dynamics and historical processes that posits whiteness, hegemonic masculinity, bourgeois values and lifestyle, heteronormativity, Christianity and Western secularism (although with many internal differences) as standards.

At the core of this book are the (pluralistic) positionality of the 'we' and its cultural roots, incorporating diverse political dynamics, social processes and geographies from Europe and its national states as well as from transatlantic ties to the rest of the West and the broader Global North. As such, the roots of the 'we' lie in a vast realm of converging and diverging cultural elements, narratives and self-narratives, which I nevertheless see as crystallising into a single voice at certain historical moments – as when homogeneous narratives and policies on migration, the war on terror and environmental catastrophes are deployed to structure the way the 'we' conceives of and responds to the threat. It is at times like those that the 'we' is made into a consolidated 'imagined community'. Just as in the wake of 9/11, a shared history and a common future are invoked to legitimise emergency measures: on the basis of a certain political and international acknowledgement of its 'common destiny', the 'we' can operate as a semiotic *dispositif* capable of developing converging strategies and common actions against *monsters* and *catastrophes*.

It is my hope that this book will contribute to understanding when and how variously assembled cultural materials feed into the discursive practices around which the identity of the 'we' solidifies. Besides *re-centring Europe and the West in the postcolonial imaginary*, the dominant narratives of our time *relieve them of all responsibility for establishing their supremacy through violence*. Present European and Western constructions of the 'we' and ideas of monstrosity and disaster will help me unpack how *white anxieties* and *moral panic* over the migrant 'invasion', 'terrorism' and the ultimate catastrophe once again reinscribe the supreme value of whiteness as the ideal standard.

Yet the focus of this work is on Europe for at least four reasons. First and more important, my politics of location compels me to read issues and events from the geographical and historical position I occupy. Second, because the Mediterranean crisis is one of the frames shaping the discursive construction of the 'we' and new humanitarian and securitarian discourses grounded in risk management measures concerned with border control and terrorism prevention. Third, because of Europe's unique geo-political proximity to areas from where both mass migration and terrorism seemingly originate – this especially applies to the southern regions along the migrant routes, owing to their geographical position on the Mediterranean. Last, because Europe is the cradle of beliefs about *history*, *geography* and *humanness* that in the last five centuries have shaped logics and ontologies of the Anthropocenic 'we'.

In contrast to traditional definitions that trace the start of the Anthropocene to European and Western Modernity, I argue that the Renaissance and Cartesianism provided the breeding ground for logics and processes that spanned the colonial endeavour, the enclosure movement, capitalist surplus-value extraction, mass enslavement and the modern patriarchal 'social contract' (Pateman 1988).

By logics of the Anthropocene, I refer to a set of principles based on ontologies of exploitation, extermination and natural resource exhaustion processes. Such principles are indiscriminately applied to organic and inorganic life through a panoply of 'technologies of power' serving the Anthropocenic order of things and determining who is worthy of benefiting from value extraction and being saved from the catastrophe (the 'we') and who is expendable. Logics and ontologies of the Anthropocene have provided a foundation for the violence of Western *history* – with its conceptions of time and progress – and *geography*, with its naturalised borders and identities, as well as for notions of *humanness* aimed at differentially including or excluding. Since colonial Modernity, the Anthropocene has been grounded in ontologies discriminating human Life from non-human (animal and vegetal) Life and Nonlife (inorganic) (Povinelli 2016) as well as in the differential attribution of humanness (Yusoff 2018). Different status and value have been assigned on the basis of those constructions of gender, race, sexuality and class that I have called 'figures of race' (Giuliani 2015a, p. 1; 2020a) – that is, intersectional constructions of race serving global and local power relations that, although unfixed and versatile, sedimented across Modernity. *Geography*, *history* and *humanness* participate in the 'common European heritage' (Mehta 1999; Lindqvist 1997) comprising 'the exterminating and genocidal colonial policies that are seen to have provided the "precursors", "incubators" and "models" for the technologies and visions that have been the cornerstones' (Stoler 2016, p. 73) of the 'we's' nation-states.

A number of voices critical of the Anthropocene narrative have been raised from within material feminist and Marxist environmental history circles, ecology movements in the Global North and the Global South, indigenous struggles and other forms of resistance at the racialised global margins. Their experiences and reflections allow me to *reveal* not only the violence integral to the workings of the Anthropocene but also the *partiality* and *situatedness* of constructions of monstrosity and catastrophe. A different take on the Anthropocene narrative is also provided by the discourses of the Capitalocene and Plantationocene. Donna J. Haraway (2015) and Jason A. Moore (2016, 2017, 2018) have argued that capitalism rather than the mere presence of human hunters, gatherers and farmers has had a massive impact on the planet through CO_2 emissions, extractivism, inhuman exploitation, genocide, air and water pollution, and land devastation. Françoise Vergès (2017), in particular, has suggested using

> the term 'racial Capitalocene' in order to emphasize the ways in which colonialism, slavery, and 'the global use of the colour line' have led to a contemporary devaluation of both human life and the nonhuman world. In understanding contemporary environmental crises, it is crucial to remain attuned to the ways in which 'destruction in the colonial era becomes visible in the postcolonial era'.
>
> (p. 77) (see Davis et al. 2019, p. 3)

The discourse of Plantationocene emphasises the role of the plantation as a system of Anthropocenic power relations and racialised bio- and necropolitics along with specific geological, ecological and biological transformations caused by colonisation, intensive plantations and enslaved labour (Tsing 2015; Haraway et al. 2015; Mitman, Haraway and Tsing 2019; Yusoff 2018; Davis et al. 2019; see also Davis and Todd 2017; Mirzoeff 2018; Moore 2015; Lewis and Maslin 2015). It is this theoretical apparatus that allows me to connect the dots between environmental catastrophes, the so-called migrant crisis, terrorism and the 'war on terror'. All are the result of what is described in scientific literature as the (neoliberal) 'acceleration', which has also led to land grabbing, mass impoverishment, devastating fossil fuel extraction, pollution, epidemics and extreme climate-change effects and, for the same reasons, to global conflicts triggered by the fossil crisis. The hegemonic view, however, is that these issues are not related. Not only the real causes are hidden and go unaddressed but also the apparatuses of governance set up by the 'we' to manage the crisis. What is needed, then, is an environmental humanities approach that looks at the relation between social, economic, political and geological dimensions of the Anthropocene. A 'more integrated and conceptually sensitive approach to environmental issues' (Bird Rose et al. 2012, p. 2) can help identify the complexly intricate causes behind the displacement of peoples and the choice to migrate that Sandro Mezzadra (2001; 2004, p. 270) calls 'objective': gender-, sexuality-, class- and race-based violence, scarcity of freedom and discrimination typical of authoritarian regimes. I also see scarcity of freedom, discrimination and structural violence as consequences of Anthropocenic logics and ontologies that distinguish between who has the 'right to have rights'

(Arendt 1951) and who does not. Not only do these 'objective causes', together with counter-terrorism measures and neoliberal practices of securitisation and differential inclusion that Europe and the West implement to mitigate the 'threat', fall within the purview of bio- and necropolitics, but they also feed economies based on and reproducing the violence of the Anthropocene.

Finally, this book sets forth *a political epistemology of care, self-care and earth-care* framed within eco-feminism, feminist 'natureculture' and social reproduction theory. Central to this epistemology is a political project capable of embracing the global crises posed by climate change, environmental disasters, migration and terrorism as well as *suturing* the historical wounds that generate monsters and catastrophes. Drawing on the anti-capitalist perspectives of degrowth and post-development movements (Kothari et al. 2019), both my feminist critique of the Anthropocene and my political epistemology challenge the neoliberal logic of acceleration, claiming the need for new concepts of 'reproduction' and care based on the necessary interdependency of *care, self-care and earth-care*. In line with Western queer (Seymour 2013), African American and African feminist theory (see Lorde 1988, p. 227; Ahmed 2014; McFadden 2020) and moving away from patriarchal and heteronormative concepts of 'care', I hypostatise self-care as limiting and substantiating care. The next step towards an understanding that self-care and earth-care are complementary and inseparable because of the interconnected materiality of the earth and the self draws on the concepts of *trans-corporeality* (Alaimo 2010) and *intra-activity* (Barad 1995, 2007, 2008) developed by material feminism to *reveal* the extreme proximity and interdependency of human Life, non-human Life and Nonlife. As human agency comes into being from intra-acting matter, *trans-corporeality* and *intra-activity* become then *political values* that, in line with feminist 'natureculture' and indigenous cosmogonies, inform new definitions of *subjectivity* and *the political*, encompassing human and non-human life, Earth and the environmental catastrophe.

3 Situating the 'we'

At the core of this work is the epistemological assumption that all knowledge production is situated, as it reflects not only the socio-historical, geographical and cultural context in which it is produced but also and more importantly the social 'location' of the producer. My epistemology is grounded in a 'politics of location' (Rich 1987; Haraway 1988) that determines 'o lugar de fala' (Kilomba 2018; Ribeiro 2019), that is, the racialised, gendered and classed position of privilege from which a particular knowledge is produced. In this case, it is the position of the 'we' as well as my own location that are under scrutiny. Both are the product of a complex set of syntactic elements and dynamics (making up their situated frames) that converge to produce both the subject of knowledge and their knowledge.

This epistemological stance arises from applying the notion of *intra-activity* to knowledge production, for which 'we know because we are of the world. We are part of the world in its differential becoming' (Barad 2007, p. 185). What this implies is that the subject of knowledge – in my case the 'we' and my 'I' – as well as their cognitive structures are the product of a pre-existing set of intra-actions among humans, and between human Life, non-human Life and Nonlife.

Also relevant to my analysis of current hegemonic narratives of monstrosity and catastrophe is the epistemological assumption that the success of any interpretation of a given moment in history depends as much on the position of privilege occupied by its author(s) as on its correspondences with pre-existing and framing conceptions. As Cyril Lionel Robert James stated referring to Herman Melville and *Moby Dick* in the last chapter of *Mariners, Renegades and Castaways* (1953, pp. 122–131) and in his famous interview with Stuart Hall (2007), any cultural work is not only the product of its context but also a collective production, since its originality depends as much on the new knowledge its author produced as on their authority to assemble existing knowledge. We could say, then, that knowledge, narratives and fears of monstrosity are genealogically produced (Foucault 1976b), as they are the result of assemblages of meanings originating from social and cultural contexts, the actors involved and the power relations existing at a certain moment.

As a historical formation regulated by a system of values, privileges, social practices and ideas, the 'we' has a specific gaze, a *right to look* (Mirzoeff 2011) legitimated by historical supremacy. As in a circular semiotic relation, this in turn legitimates all the distinctions that gave substance to its political, economic and cultural supremacy in the past. The same distinctions – e.g. between the 'we' and the 'Other', the Here and the Out there, the 'good' and the 'bad', the 'worthy' and the 'expendable' – operate today in narratives of monstrosity and fears of catastrophe. Conceptions of the Out there as the place of the 'expendable' are as relevant today as in the past and deeply entrenched within modern perceptions of a divide between the West and the rest of the world and the feeling of being 'under siege' (Hage 2016). Since early Modernity, ideas of *monsters* and *catastrophes* as external beings and events have been reinforced by the assumption that Europe and the West are isolated, under attack and faced with a threat coming from what is perceived to be an *incommensurable* and *ontologically dangerous* outside, a place of darkness (Conrad 1899). Although differently signified over the course of time as the place of unmasterable nature, systemic disasters, infidels and indigenous monsters, the Out there has always been identified with the unknown threatening Otherness.

If the space of the 'we' is considered to be free from disaster (by virtue of civilisation having mastered nature, reducing it to a repository of resources for value extraction), it follows that the Out there is the 'place *for* disaster'. Catastrophe, as opposed to disaster, has been viewed since the *siècle des*

Lumières as a *crisis*, a time for *renewal* inaugurating a new phase in the history of humanity. In late Modernity, as Benjamin argued (1940), humans did not let catastrophe dim their hopes for the future. According to environmental humanist scholar Miriam Tola, this attitude was a reflection of the 'Anthropocene master narrative in which an undifferentiated Anthropos is simultaneously the cause of catastrophe and its remedy' (Tola, personal communication, 8 March 2020). In this narrative, which today permeates ecomodernism,[1] catastrophe was a mark of the 'civilised space' of the West, while disasters were viewed as structural destructive events and, as such, identified with the Out there. This Anthropocenic belief fed into colonial Modernity (Said 1978, 1993) reproducing Western ideas of the Out there as inherently chaotic, dangerous, violent, backward and lacking risk-predicting and crisis management expertise. As argued by Edward Said in his description of the Orient in the British imperial imagination (1978), and more recently by Achille Mbembe about Africa and Africans in Western postcolonial discourses (2001, p. 9), the monstrous Other and their Out there have been construed as being responsible for their own apocalypse.

The Other, be they coming from a geographically severed or internally segregated Out there, is always a threat in the eyes of the 'we' – an agent of corruption that may ultimately lead to the annihilation of the 'body' of the nation, Europe or the West. Migrants arriving on Lampedusa, Malta or Lesbos (Giuliani 2016b) are blamed for gradually forcing Europe's economic and social collapse, spreading deadly diseases and plotting terror attacks. The Other and the Out there are *risky*, by definition (Aradau 2004; Salerno 2016, p. 374).

As Judith Butler argued, referring to changing conditions and perceptions in the United States and the West after 9/11, the definition of *risk* today is part and parcel of the *revelation* of our 'fundamental dependency on anonymous others [...]. No security measure will foreclose this dependency; no violent act of sovereignty will rid the world of this fact'. And although 'there are ways', as Butler adds, 'of distributing vulnerability, differential forms of allocation that make some populations more subject to arbitrary violence than others' (2004, p. xii), there has been to this day no sufficient public and political *awareness* of the fact that violence and reciprocation necessarily hit, first and massively, the most vulnerable. Neither has there been institutional acknowledgment of the questions raised by Sara Ahmed in 2004, 'Who is contained through terror? Whose vulnerability is at stake?', by which she clearly pointed at the differential economy of 'fear' and 'anxiety' grounded in postcolonial power relations based on race and gender.

The context in which the 'frame of war' was introduced after 9/11 is that which Ulrich Beck, back in the early nineties, called the *risk society*, in which the 'new modernity' of globalisation makes the life of the privileged vulnerable and precarious, exposed as it is to unpredictable events (1992, p. 13).

What Beck refers to is a new 'acceleration' – also acknowledged by scientific literature on the Anthropocene – putting tremendous strain on nature's ability to contain the risks inherent in a globalised modernity. These are not 'side-effects', he continues, but outcomes themselves of the post-industrial *risk society*. Beck's intuition would be correct, if it were not for the fact that not all societies reached the post-industrial stage at the same time and in the same way, and some have not reached it at all. His argument assumes a universality that does not exist and a common vulnerability to the effects of global capitalism that many 'Others' have actually already experienced. Who is becoming vulnerable? Who has always been? At any rate, 30 years later, mainstream narratives and official accounts of climate-change impacts and environmental catastrophes still refer to the risks posed by the last neoliberal acceleration of the Anthropocene as 'side-effects'. The depersonalisation of responsibility casts the Anthropocene as a product of 'human nature' as a whole, while it is, instead, a set of 'interrelated historical processes set in motion by a small minority' (Davis et al., 2019, p. 4).

Blanket blaming humanity as a whole, with no distinction of race, gender, class or nationality, has led to a general lack of accountability as well as to the depoliticisation and gradual normalisation of the crisis (Bettini 2013), as a result of which protests are silenced and those who flee disaster are criminalised, turned into *monsters* who do not accept a life of subjection and sacrifice and who cross walls, borders and segregated spaces to reach the (more) 'protected' space of the 'we', endangering it. In this frame, it is those who have benefited the least from value extraction and extractivism who would be sacrificed first in the event of an apocalypse that is being presented as nothing more than a crisis, a period of transition to a post-apocalyptic world in which logics and ontologies of the Anthropocene will once again prevail. Today as in the modern colonial past, racialised constructions of monstrosity and catastrophe are used as *dispositifs* to ostracise those who are perceived as threatening (Foucault 1976b; Nuzzo 2013, p. 56) and divert attention away from the *real* monster.

4 Monstrosity, race and the archives

The key assumption behind *Monsters, Catastrophes and the Anthropocene: A Postcolonial Critique* is that the symbolic material composing the figures of *monster* and *monstrous catastrophe* has sedimented across colonial and postcolonial temporalities and spaces (Kabbani 1986; Mercer 1986; Fredrikson 1971; Hancock 1988; Yeğenoğlu 1998; Hobson 2005; Bell 2018). The 'figures of race' (Giuliani 2015a, p. 1) are deeply entrenched in today's vast cultural production around disasters and crises – be it about alien invasions, natural disasters, global terror attacks, zombie uprisings or the ever-increasing swathes of migrants reaching European shores (Giuliani 2017; Giuliani 2020b; Bhatia et al. 2018; Agier 2018). All contribute to the production of a *hyperreal* threat based on a racialised, gendered and classed dichotomy between the worthy 'we' and the expendable Other. The thousands of refugees who peacefully walked across Europe in hopes of finding asylum in 2015 were also considered a threat. Discourses and visual narratives of desperate masses penetrating the 'safe' space of the 'we' contributed to a widespread perception of racialised chaos that culminated in the *moral panic* over refugees sexually assaulting women later that year in Cologne (Garraio 2019).

In colonial Modernity, any subversion of the racialised Order was represented as monstrous, and monsters as violent and deadly (Scarry 1985, p. 33; McClintock 1995; Hartman 1997, p. 51; Sharpe 2009; Giuliani 2020b). Appropriation of colonial monstrous figures in the metropole, and vice versa, generated transnational articulations that were culturally and contextually adapted to local power relations. Racialised figurations of the infidel harking back to the legal terminology developed for the Crusades were used in descriptions of the indigenous peoples during the Spanish conquest of the Americas (Nuzzo 2004). Likewise, racialised and gendered figures of enslaved people from plantation-based societies in North America and elsewhere were rearticulated and applied to post-colonial relationships with Africa (Giuliani 2016b, p. 94), while images coined in the United States, Australia, Canada, and East and South Africa to describe Asian migrants and contract labourers (the so-called coolie trade) were borrowed in Europe, where migration from Asia started at a later time and in a much different guise (Giuliani 2016c).

In line with Stuart Hall, I consider race a 'textual thing' that is not fixed in its inner nature, 'cannot be secured in its meaning', and 'floats in a sea of relational differences'. 'After all the violence, [it] remains a social construct; it changes with the context, it is relational, not essential, and [it can] never be finely fixed, but [is] subject to the constant process of redefinition and appropriation' (1997) [*transcribed by the author*]. Race is a 'signifier' whose meanings rely on what Gloria Wekker and Ann Laura Stoler, in line with Edward Said (1978), call, respectively, (national) cultural and colonial archives. Colonial and (national) cultural archives are 'sites of knowledge production' (Stoler 2002, p. 90), 'repositor[ies] of codified beliefs' (p. 97) that have 'influenced historical cultural configurations and current dominant and cherished self-representations and culture' (Wekker 2016, p. 2). In general, colonial and (national) cultural archives have both local and transnational origins and produce different knowledge depending on the reader, the time in history, the social contexts and the power relations they serve. On the other hand, the modern racialisation of the colonial subject developed in constant exchange with that of the internal abject (women as well as cultural and religious minorities), while their dehumanisation was accomplished through converging processes of signification that included animalisation and the attribution of moral depravity (Friedman 1981; Creed 1993; Hobson 2005; Jeffreys 2005; Sharpe 2009; Baker 2010). As shown, for instance, by historian Silvia Federici (2004), art historian

John Block Friedman (1981), historian of philosophy Charles Wolfe (2005) and political philosopher Filippo del Lucchese (2019), monstrification was based on transcendent and secular conceptions associating the lower order of living things (that is, animals, perceived as farthest from the light and beauty of God) with cannibalism, hybridity, non-humanness and horrific mutations.

As a relational category, race is differentially constructed on the basis of the meanings assigned to hegemonic whiteness (Guillaumin 1972), which is not a fixed construct either but changes across time depending on values, narratives and social components. The textuality of race and whiteness has material effects on people's lives and differentially affects the structures and functioning of society, state and market: thus just like the 'figures of race', race and whiteness are semiotic 'technologies of power' (Foucault 1976a) whose cultural nature (Young 1990) has never prevented them from producing material violence. Race and whiteness are also powerful ideological apparatuses. Their ideology 'materialises' so-called racial differences, turning them into biological or genetic 'truths' based on the racialised subject's allegedly serial and predictable behaviours, which sustain the lasting dominance of the figures of race and the 'we' they contribute to creating. Figures of race, including figures of whiteness, may be seen as a derivate semiotic *dispositif* of *race* – figurations stored in the cultural archives of the West and which are in a circular relation with the entity (being it individuals, societies or institutions) that activates them, the object of racialisation and the power relations they serve. When the 'we' comes together against a perceived threat, that is, when capitalist power structures of gender and race seem to backfire on Europe and the West, the 'figures of race' are rapidly activated to provide a reliable explanation of the threat and justify protective measures. In the face of environmental catastrophe, Anthropocenic ontologies and logics taken to the extreme have led to a neoliberal 'acceleration' (Beck 1992) of capitalism that is deeply affecting the circular semiotics of race through the so-called multiplication of borders (Mezzadra and Neilson 2013), the securitisation of migration (Huysmans 2006), the role of 'humanitarian reason' in strengthening the border regime (Fassin 2011; Aas and Gundhus 2015) and the narrative of terrorism reinforcing the modern dualism of civilisation and barbarity (Amoore and De Goede 2008). In this context, race determines an individual's degree of vulnerability and the 'figures of race' provide a legitimising dispositif which appears inextricably bound to the rapid mediatisation of monsters and catastrophes. Today, iconographies of race are central to the mobilisation of *moral panic*.

5 The role of visual culture

What lies behind the modern imaginary of monsters and catastrophe is a 'set of signifying practices and figures from the archive of white Western popular culture' (Demaria 2016, p. 143, my translation) that connect culture and materiality, presenting past memories, present fears and future expectations as 'universal'. *Reality* and media narratives appear to conflate into *hyperreality*, producing a biased narrative of the threat that is construed as 'objective' and true for everyone, despite advancing the idea that the 'we' is the only entity worth protecting. Visual materials are part and parcel of the discursive fabric fuelling the 'we's' *moral panic* over a threat (Cohen 1972; Hall et al. 1978; Bhatia et al. 2018; Graham 2002) that, as Sofia José Santos argues:

> is framed in such a way as to be interpreted and perceived as a disruption of the social cohesion and the social order, thus assuming a moral dimension. The construction of these security narratives does not occur, nonetheless, instinctively, nor does it emerge within a discursive and ideologically unfilled context (Barrinha 2011). Rather, it is a process that is rooted and sustained upon a broader ideological, historical, power-driven, and identitarian discursive structure – where gender, nation, race and class perform key roles – that shapes and validates it (Santos et al. 2018). In this context, narratives and messages are constructed in order to have a compelling and mobilizing meaning to their specific audiences.
>
> (Santos in Giuliani et al. 2020a, p. 165)

My argument is that, despite internal differences, converging discourses in Western mainstream visual culture emphasise the role of the 'we' and Here as the bastions of human rationality and techno-scientific progress to be preserved at the expense of the Other.

Key differences in the way film, television and the news media frame *moral panic* owe to the fact that the primary purpose of popular culture is entertainment. Apocalyptic narratives make the most of doomsday scenarios and horrific monsters to amplify *moral panic*, whereas the news media are expected to exercise ethical restraint – although they are just as keen on attracting audiences through images and discourses that belong to the cultural repertoire of everyday life (see also Dayan and Katz 1994; Liebes and Curran 1998; Cottle 2006). Forms of *transmediality* and *intericonicity* (Chéroux 2007) allow media and non-media iconographies to be circulated across multiple formats, likely exerting significant influence on information processing and retention as well as shaping public attitudes. Mainstream media images of migration, for instance, repeatedly portray the 'Other-at-sea' as distant, not belonging,[2] *risky* and *at-risk*, both threatening and 'reassuring' in their vulnerability[3] (Giubilaro 2018, p. 12; Salerno 2016; Giuliani 2017; Pogliano and Solaroli 2012; Jacomella 2015; Pogliano 2019; Giuliani and Vacchiano 2019; Santos et al. 2018).[4] Migrant landings on Europe's southern shores are often represented as *alien* invasions requiring military force (as in the case of the Hungarian–Serbian, Turkish-Greek and Croatian-Bosnian Herzegovinian borders).[5] This same iconography is allegorically rendered in apocalyptic films, where the figures of the hungry, angry and bloodthirsty terrorist alien and rabid monster (the *fast running and devouring rabid undead*) are familiar to audience members from representations in mainstream visual and print media (Santos et al. 2018; Giuliani et al. 2020a). Moreover, widely circulated images of the environmental catastrophe and mainstream narratives of the Anthropocene[6] populate media and film narratives in which a planetary catastrophe is brought about by a planet-to-planet collision, the sun's cooling, climate change, epidemics, or human genetic mutation.

This does not mean that all trans-medial and intericonic material will necessarily reproduce a biased imaginary. My method offers a rather selective, if not reductive, reading of visual products, which, although it may not do justice to their textual complexity, serves the purpose of deconstructing the 'we'/'they' *dispositif*. Films and TV series on zombies, aliens, and environmental apocalypses broadcast in cinemas and online platforms were selected on the basis of my research interest in the modern genealogy of fears and anxieties experienced by the 'we'. Besides *trans-mediality* and *intericonicity*, other criteria guided my selection: the repeated use of images of monstrosity and catastrophe, the themes addressed and their connections to current philosophical debates on human mobility, global security and the Anthropocene. Whether apologetic or critical of mainstream understandings of the Anthropocene and planetary catastrophe, they are an expression of the cultural milieu within which the imaginary of catastrophe and the 'After' exists. In any case, such *analysis* is *not undertaken* simply for *its own sake*. Films and TV series are neither case studies nor texts encapsulating some hypothetical truth. On the contrary, they offer philosophical elements that I combine with theoretical insights to develop my argument.

My interest in them stems from their being 'memories of the future', predictions of what the future will be like based on collective memories of the past. Just as memories of the past are the result of processes of selection and reassessment mediated by social and individual factors, so too symbolic material, perceptions, fears and fantasies of the future mediated by understandings of the present and its imminent threats recombine to form 'memories of the future'. Seen from another perspective, they elaborate on the future through allegories of the past, projecting onto futuristic reassuring or dystopian 'after-worlds' fears and hopes of the present time. In his philosophical analysis of two films that I also consider here, *28 Days Later* (2003) and *28 Weeks Later* (2007), Anirban Kapil Baishya (2011) writes that:

> post-apocalyptic horror and science fiction cinema taps into the *optical unconsciousness* [Benjamin 1931] to refer back to the horrors of catastrophic events, by evoking the images of war, violence and destruction and reproducing the tortured corporeality thereof by means of oblique reference. This method and moment of reference is what Lowenstein (2005, pp. 1–16) calls the 'allegorical moment'.
>
> (p. 4)

Memories of the future are extremely important to my analysis, as they signal hegemonic and critical views of the past and the present (Giuliani 2015b, 2016a, 2020). They capture the essence of what I call the 'repressed memories' of Anthropocenic modern and colonial violence – institutionally revised, silenced, or collectively denied. Further, they include dystopian plans for survival that are only apparently futuristic but not really so, as they seem to have an uncanny correspondence with the real world. Three such examples are found in *Code 46* (2003), *Elysium* (2013) and the online TV series *3%* (2016–), which envision new colonies, segregated spaces and carceral archipelagos that bear some resemblance to present-day *dispositifs*. The ability to construe *revealing* 'allegories' of the present, as Roger Luckhurst maintains, turns science fiction, fantasy and horror into '"planetary fictions", uniquely aware of the imbrication of story and world in a way that more domesticated realism cannot begin to grasp' (2015, p. 184). In brief, their 'allegories' can even more openly than realistic narratives show:

> how the uneven sedimentations of colonial reason and the effective sensibilities on which they depend – whether under the rubrics of 'security,' 'terrorism,' 'defence of society', [sic] or 'race' – participate in shaping the possibilities for how *differential futures* are distributed and who are, and will be, targeted as those to be exposed, both external and internal enemies in the making.
>
> (Stoler 2016, p. 13)

6 The book's contents

This book comprises three main sections. **Section 1** focuses on constructions of monstrosity and catastrophe in the context of post-9/11 migrations towards Europe and the West. Its aim is to challenge mainstream descriptions of migration as an 'invasion' through an inquiry into the symbolic materials that contribute to such representations creating *moral panic*. The figure of the fast running and devouring rabid zombie will help unveil *what remains of the modern colonial cannibal* today in the 'we's' imaginary of human mobility from the Global South. More specifically, it will contribute to an understanding of how its corresponding 'figures of race' in current descriptions of migration serve the purpose of *re-centring Europe and the West in the postcolonial imaginary and power relations*. The genealogy of the *cannibalistic monster* is framed into a conception of *history* as an epistemological battlefield, in which bio- and necropolitics implemented to control, channel and filter human mobility are publicly disassociated with similar practices implemented in the past to subject the racialised Other to colonial rule or slavery (Saucier and Wood 2014; Danewid 2017; Casid 2018). It is within this frame that the postcolonial '*fortressed*' West is re-signified today as the last bastion of reason, justice and humanness against the overflowing of barbarity into the world of the 'we'. In this case, the semiotic power of the border – a concept that I introduce in this section and cover in greater detail in **Section 2** – contributes to constructing the 'we' as the righteous 'victim' of barbarity and yet providing shelter for those in need of benevolent assistance (i.e. migrants and refugees).

The *fast running and devouring rabid undead* is a staple of films such as *28 Days Later* (2003) by Danny Boyle, its sequel *28 Weeks Later* (2007) by Juan C. Fresnadillo, Francis Lawrence's *I am Legend* (2007), *The Horde* (2009) by Yannick Dahan and Benjamin Rocher, Marc Forster's *World War Z* (2013) and TV series like the BBC's *In the Flesh* (2013–2014) by Dominic Mitchell and Jonny Campbell (2013–2014), and Frank Darabont's *The Walking Dead* (2010–). Cinematic zombies, the subject of many a horror film since their first appearance in *28 Days Later*, epitomise fears that a history of horrors will strike back. Traditionally, their roots have been traced back to the voodoo undead in Afro-Caribbean and Latin American slave culture. However, I believe these figures have a significance that extends well beyond transatlantic and transpacific slave and colonial practices and cultures, in that they reflect publicly repressed, troublesome memories of horrific regimes of segregation, sexual violence, forced reproduction and death also shared by both victims and perpetrators in Europe and the West as a whole. For this reason, I will trace the genealogy of the anthropophagic monster in literature, political theory and popular visual culture from the Middle Ages to the present, reading today's migrant boat and the transatlantic slave ship contrapuntally. The migrant boat crossing the Mediterranean emerges from my reading as a reverse *chronotrope* of the slave ship, allowing me to connect *white anxiety* and *moral panic* to the autonomy of migration. In fact, while *moral panic* in the colonial and slavery systems was caused by the racialised subject violating containment measures, *moral panic* and the monstrification of border trespassing today are a reaction to the autonomy of migration from the Global South. Through the selective use of the figure of the *fast running and devouring rabid monsters* – which also operates as an allegory of the speed and anthropophagic nature of neoliberal capitalism – I will contrast the panic generated by unruly hordes of rabid cannibals with the panic generated by border trespassing, sometimes described in the news and in public debates as an overwhelming epidemic. My analysis of the gendered, racialised and classed fantasies behind narratives of zombie uprisings, human catastrophes and the 'after' will end the section, allowing an understanding of the positionality of the gaze through which the catastrophe is imagined and the 'worthy' are identified.

Section 2 focuses on the process of monstrification of border trespassing. After tracing a genealogy of constructions of monsters in motion – rebels transgressing the order of all sorts of borders, camps and racially segregated colonies – I will connect them to current representations of autonomous migration. In this section, I will analyse seven alien films: *Monsters* (2010) by Gareth Edwards, *Under the Skin* (2013) by Jonathan Glazer, *District 9* (2009) by Neill Blomkamp, *War of the Worlds* (2005) by Steven

Spielberg, *The Mist* (2007) by Frank Darabont, *Arrival* (2016) by Denis Villeneuve and *Annihilation* (2018) by Alex Garland. Through the figure of the alien invader and the survival strategies and *dispositifs* devised in film and TV narratives, I will enquire into postcolonial fears of reverse invasion (Hage 2016) and colonisation. Such fears project onto the alleged invasion all forms of violence, enslavement and genocide that the 'we' perpetrated on its colonial peripheries and subjects. That modern and colonial *geography* will vanish, together with the boundaries and borders that have spatially separated and protected the 'we' from the Other, is one feared outcome of this invasion. Barbarity and catastrophe are believed to be just around the corner, eventually transforming the safe space of the 'we' into a place *for* disaster. The figure of the alien invader allows me to delve into the workings of modern bio- and necropolitical *dispositifs* that monstrify the threat by distancing it from society and the 'we'. These same spatial *dispositifs* project monstrosity and chaos across the boundaries of the 'we' – that is, onto the Out there, the Outside and the Otherworldly. As a result, the Here and the Out there are not only physically distanced but also assigned opposite moral ontologies reflecting the opposition between Good and Evil. Among the spatial *dispositifs* that throughout Modernity maintained this dichotomy are the border (Balibar 2002; De Genova 2013; Mai 2014; Tazzioli 2016, 2017a, 2017b; Tazzioli and Walters 2016, 2019), the camp, the colony, the carceral archipelago (Agamben 1995; Stoler 2016) and the wall (Brown 2010). I will examine their semiotic power to reactivate colonial and cultural archives and 'figures of race': sustained by pre-existing constructions of race, these *dispositifs* reproduce them rearticulating their meanings to serve the specific context of neoliberal border regimes. Drawing on the debate within critical studies regarding security and terrorism, I will explore which constructions of race sustain the legitimacy of practices of segregation, encampment, quarantine, exclusion and differential inclusion, at a time when the border is feared to be 'mobile' and 'incorporated' (Bodemann and Yurdakul 2006) by the internal enemy (the alleged sleeper terrorist) and the migrant/refugee invader.

Alien films present racial difference as a mark of absolute Otherness projected onto the alien – who is alternately seen as embodying a deadly *terrorist*, and hence the object of old and new forms of segregation, or as coexisting with the 'we' in new and old border-zones. Some contrast dystopian visions of walled communities and segregated aliens with utopian *skin-to-skin* co-habiting (Ahmed 2000). In others, there is room left for hope of inter-speciesist Chthulucenic communication (Haraway 2016). Worldly–otherworldly communication takes the form of either a logical exchange through translation and language sharing (*Arrival)* or a *bio*logical exchange through DNA and radio wave refraction (*Annihilation*). Both imply the existence of a sort of *partage*, or *skin-to-skin* proximity, between human and non-human Life (Iveković 2019; Ahmed 2000), allowing for a post-humanist exchange to take place.

Finally, **Section 3** explores how environmental and climate-change catastrophes *reveal* the *monstrosity* of the Anthropocene – its ontologies and logics. By revealing the unacknowledged obscene violence that preserves the anthropocentric order, catastrophe challenges modern dualisms of culture/nature, human/animal, human/non-human, organic/inorganic, 'here'/'out there', civilised/barbarian, centre/peripheries of the world that sustain the workings of the Anthropocene. Nevertheless, in the face of a catastrophe that will destroy all borders, walls and segregated spaces, freeing 'barbarity' from its constraints, the 'we' still asks for more Anthropocene – and to persist with the monstrification of the vulnerable and the Out there. Among the films and TV series I take into account in this section are *The Core* (2003) by Jon Amiel, *Code 46* (2003) by Michael Winterbottom, *The Day After Tomorrow* (2004) by Roland Emmerich, *Blindness* (2008) by Fernando Mereilles, *The Impossible* (2011) by Juan Antonio Bayona, *Snowpiercer* (2013) by Boong Joon Ho, *Elysium* (2013) by Neill Blomkamp, *The Girl with All the Gifts* (2016) by Colm McCarthy, *I Am Mother* (2019) by Grant Sputore, *The Wandering Earth* (2019) by Frant Gwo and the web TV series *3%* (2016–) by Pedro Aguilera.

A historical reconstruction of the genealogy of meanings associated with the environmental catastrophe in pre-modern and modern Western culture allows me to explore the turning point represented by the *siècle des Lumières* and ideas of disaster as an opportunity for social *renewal*. What I argue is that, despite the planetary nature of both climate change and the ongoing environmental catastrophe, post-apocalyptic scenarios in scientific and popular culture are still marked by the presence of the Anthropocenic Anthropos enacting his *renewal*. A future without humanity seems unthinkable for a number of reasons, including the unwillingness of the 'we' to come to terms with the wrongs of the Anthropocene. In the face of the vanishing Anthropocenic *geography* and *history*, signalled today also by mass migration into the safe space of the 'we' and the threat of terrorism, humankind seeks solace in the thought that the environmental catastrophe and its effects on the world may well be limited in space and time. This scenario is a staple of doomsday films and TV series: all but one of the films in this section see humans reproducing Anthropocenic fantasies of selection, colonisation and slavery in the post-apocalyptic world. Some (*Code 46*, *Blindness*, *Snowpiercer*, *Elysium*, *I Am Mother* and *3%*) depict a dystopian 'after', denouncing the reactivation of fantasies of Malthusianism, social Darwinism and eugenics as the means for preserving 'human society'. Nevertheless, none dare to critique the capitalist logic of development and growth. In the European and Western popular and hegemonic imaginary of catastrophe, the blame is on the irresponsible conduct of a few, and all that is needed to avoid it is minor tweaks to capitalism.

Viewing the existing interdisciplinary debate (Barca 2020) on degrowth, 'natureculture' and social reproduction through a postcolonial feminist lens, I argue that *the political* needs to find a new foundation in assumptions of *trans-corporeality*, *intra-activity* and interdependency of Life and Nonlife – a political project based on the interdependency of care, self-care and earth-care. The British film *The Girl with All the Gifts* (2016) will provide me with a background for introducing my idea of *trans-corporeality* and *intra-activity* as new foundational political values. The figure of human hybrid, in particular, will support my argument that recognition of the political *generativity* of *intra-activity* and interdependency of care, earth-care and self-care need to replace the ideal of 'a pure and pristine' humanness.

In the conclusion, I will sum up my analysis: its first part, 'COVID-19: Chronicles of the "we" in the time of the pandemic', will explore the imposition of bio- and necropolitics in Europe and the West during the 2019–2020 coronavirus pandemic through news articles, text messages, Facebook posts and private messages from friends, relatives and colleagues. Drawing parallels between the entries in the chronicle and the memories of the future that I have discussed in this book, I will highlight the striking similarities between the bio- and necropolitical apparatuses described in films and TV series and the measures taken to contain chaos in the space of the 'we'. My reflections will stress the Anthropocenic nature of the crisis caused by a zoonotic spill-over that can be attributed to intensive meat production, the sale of live animals at wet markets, and disruptions to ecosystems caused by rapid urbanisation.[7] In its second part, 'The end of the word as the "we" knows it?', taking a stance against logics of surplus-value extraction from Life and Nonlife, I will present the actual implementation of my anti-Anthropocenic political project: strategies of sharing, caring, self-caring and earth-caring in the locked-down world of the 'we' provide the blueprint for a collective rethinking of the 'Day After' the Anthropocene.

Notes

1 www.ecomodernism.org/.
2 www.tdg.ch/monde/europe/L-UE-se-penche-sur-sa-politique-dimmigration/story/15511242.
3 www.dailystar.com.lb/News/World/2019/Mar-06/478143-more-than-80-migrants-rescued-off-italys-lampedusa.ashx.
4 www.dailymail.co.uk/news/article-1371204/Lampedusa-MORE-migrants-fleeing-Tunisia-Libya-inhabitants.html.
5 https://foreignpolicy.com/2019/12/06/croatia-is-abusing-migrants-while-the-eu-turns-a-blind-eye/; www.aljazeera.com/news/2020/01/croatia-police-criticised-attacking-refugees-border-200115105827903.html;www.reuters.com/article/us-europe-migrants-hungary-police/two-hungarian-police-officers-fined-for-brutality-against-migrants-newspaper-idUSKBN16N16I.
6 For example, those presented at Edward Burtynsky, Jennifer Baichwal and Nicholas de Pencier's exhibition 'Anthropocene' at the MAST in Bologna, which I attended in January 2020 https://anthropocene.mast.org/.
7 www.greenqueen.com.hk/preventing-another-pandemic-the-link-between-coronavirus-and-industrial-livestock-farming/.

References

Aas, K. F. and Gundhus, H. O. I. (2015) 'Policing humanitarian borderlands: Frontex, human rights and the precariousness of life'. *British Journal of Criminology* 55 (1), pp. 1–18. https://doi.org/10.1093/bjc/azu086

Agamben, G. (1995) *Homo Sacer: Sovereign Power and Bare Life*. Stanford: Stanford University Press (this edition 1998).

Agier, M. (2018) *The Jungle: Calais's Camps and Migrants*. Cambridge: Polity Press.

Ahmed, S. (2000) *Strange Encounters*. Abingdon and New York: Routledge.

Ahmed, S. (2004) *The Cultural Politics of Emotions*. Edinburgh: Edinburgh University Press. http://doi.org/10.4324/9780203700372

Ahmed, S. (2014) 'Selfcare as warfare'. *Feminist Killjoy*. [Blog] Available at: https://feministkilljoys.com/2014/08/25/selfcare-as-warfare/

Alaimo, S. (2010) *Bodily Natures: Science, Environment, and the Material Self*. Bloomington: Indiana University Press.

Amoore, L. and De Goede, M. (eds) (2008) *Risk and the War on Terror*. New York: Routledge. https://doi.org/10.4324/9780203927700

Anderson, B. (1983) *Imagined Communities: Reflections on the Origin and Spread of Nationalism*. New York: Verso (this edition 2006).

Aradau, C. (2004) 'The perverse politics of four-letter words: risk and pity in the securitisation of human trafficking'. *Millennium: Journal of International Studies* 33 (2), pp. 251–277. https://doi.org/10.1177/03058298040330020101

Arendt, H. (1951) *The Origins of Totalitarianism*. Orlando, Austin, New York, San Diego and London: Harvest Book (this edition 1973).

Asad, T. (2007) *On Suicide Bombing*. New York, Columbia University Press.

Baishya, A. K. (2011) 'Trauma, post-apocalyptic science fiction and the post-human'. *Wide Screen* 3 (1), pp. 1–25.

Baker, N. (2010) *Plain Ugly: The Unattractive Body in Early Modern Culture*. Manchester and New York: Manchester University Press. http://doi.org/10.2307/j.ctv6p50z

Balibar, É. (2002). *Politics and the Other Scene*. London: Verso.

Barad, K. (1995) 'Meeting the universe halfway. Ambiguities, discontinuities, quantum subjects, and multiple positioning in feminism and physics'. In: Hankinson Nelson, L. and Nelson, J. (eds) *Feminism, Science and the Philosophy of Science*. Norwell: Kluwer Press, pp. 161–194.

Barad, K. (2008) 'Posthumanist performativity: toward an understanding of how matter comes to matter'. In: Alaimo, S. and Hekman, S. (eds) *Material Feminisms*. Indianapolis: Indiana University Press, pp. 120–136.

Barad, K. (2007) *Meeting the Universe Halfway: Quantum Physics and the Entanglement of Matter and Meaning*. Durham, Duke University Press. http://doi.org/10.1215/9780822388128

Barca, S. (2020) *Forces of Reproduction. Notes for a counter-hegemonic Anthropocene*. Cambridge: Cambridge University Press.

Barrinha, A. (2011) 'The political importance of labelling: terrorism and Turkey's discourse on the PKK'. *Critical Studies on Terrorism* 4 (2), pp. 163–180. https://doi.org/10.1080/17539153.2011.586203

Beck, U. (1992) *Risk Society: Towards a New Modernity*. London: Sage.

Bell, D. (2018) *Globalizing Race: Antisemitism and Empire in French and European Culture*. Chicago: Northwestern University Press. http://doi.org/10.2307/j.ctv47wbp5.

Benjamin, W. (1931) 'Little history of photography'. In: Benjamin, W. *Selected Writings. Volume 2 (1927–1934).* London, MA: The Belknap Press of Harvard University Press, pp. 507–530 (this edition 1999).

Benjamin, W. (1940) 'Theses on the philosophy of history'. In: Benjamin, W. *Illuminations: Essays and Reflections.* New York: Schocken Books, pp. 253–264 (this edition 1968).

Bettini, G. (2013) 'Climate barbarians at the gate? A critique of apocalyptic narratives on "climate refugees"'. *Geoforum* 45, pp. 63–72. https://doi.org/10.1016/j.geoforum.2012.09.009

Bhambra, G. (2015) 'Europe won't resolve the "migrant crisis" until it faces its own past'. *The Conversation.* http://theconversation.com/europe-wont-resolve-themigrant-crisis-until-it-faces-its-own-past-46555. Accessed 15 October 2019.

Bhatia, M., Poynting, S. and Tufail, W. (eds) (2018) 'Introduction'. In: *Media, Crime and Racism.* Basingstoke: Palgrave. http://doi.org/10.1007/978-3-319-71776-0

Bird Rose, D., van Dooren, T., Chrulew, M., Cooke, S., Kearnes, M. and O'Gormand, E. (2012) 'Thinking through the environment, unsettling the humanities'. *Environmental Humanities* 1, pp. 1–5. http://doi.org/10.14361/transcript.9783839421352.fm

Bodemann, M. Y. and Yurdakul, G. (eds) (2006) *Migration, Citizenship, Ethnos: Incorporation Regimes in Western Europe and North America.* New York: Palgrave Macmillan. https://doi.org/10.1057/9781403984678

Braidotti, R. (2013) *Post-human.* Cambridge: Polity Press.

Brown, W. (2010) *Walled States, Waning Sovereignty.* New York: Zome Books.

Butler, J. (2004) *Precarious Life: The Power of Mourning And Violence.* New York: Verso.

Butler, J. (2009) *Frames of War. When Is Life Grievable?* New York: Verso.

Casid, J. H. (2018) 'Necropolitics at sea'. In: Gualtieri, C. (ed.) *Migration and the Contemporary Mediterranean. Shifting Cultures in Twenty-First Century Italy and Beyond.* Oxford, Bern, Berlin, Brussels, New York and Wien: Peter Lang, pp. 193–214.

Chéroux, C. (2007) 'Le déjà-vu du 11-Septembre. Essai d'intericonicité'. *Études photographiques* 20 (June), pp. 148–173.

Clark, N. (2011) *Inhuman Nature: Sociable Life on a Dynamic Planet.* London: Sage. http://doi.org/10.4135/9781446250334

Cohen, S. (1972) *Folk Devils and Moral Panics.* London: MacGibbon and Kee.

Conrad, J. (1899) *Heart of Darkness.* New York: Tribeca Books (this edition 2010).

Cottle, S. (2006) 'Mediatized rituals: beyond manufacturing consent'. *Media, Culture & Society* 28 (3), pp. 411–432. https://doi.org/10.1177/0163443706062910

Creed, B. (1993) *The Monstrous-Feminine: Film, Feminism, Psychoanalysis.* London: Routledge. https://doi.org/10.4324/9780203820513

Danewid, I. (2017) 'White innocence in the Black Mediterranean: hospitality and the erasure of history'. *Third World Quarterly* 38 (7), pp. 1674–1689. https://doi.org/10.1080/01436597.2017.1331123

Davis, H. and Todd, Z. (2017) 'On the importance of a date, or decolonizing the Anthropocene'. *ACME: An International Journal for Critical Geographies* 16 (4), pp. 761–780.

Davis, J., Moulton, A. A., Van Sant, L. and Williams, B. (2019) 'Anthropocene, capitalocene, … plantationocene?: a manifesto for ecological justice in an age of global crises'. *Geography Compass* 13 (5), e12438. https://doi.org/10.1111/gec3.12438

Dayan, D. and Katz, E. (1994) *Media Events.* Cambridge, Mass. and London: Harvard University Press.

De Genova, N. (2013) 'Spectacles of migrant 'illegality': the scene of exclusion, the obscene of inclusion'. *Ethnic and Racial Studies* 36 (7), pp. 1180–1198. https://doi.org/10.1080/01419870.2013.783710

Del Lucchese, F. (2019) *Monstrosity and Philosophy: Radical Otherness in Greek and Latin Culture*. Edinburgh: Edinburgh University Press.

Demaria, C. (2016) 'Postfazione'. In: Giuliani, G. *Zombie, alieni e mutant. Le paure dall'11 settembre ad oggi*. Florence-Milan: Le Monnier/Mondadori Education, pp. 143–153.

Esposito, R. (2008) *Bíos: Biopolitics and Philosophy*. Minneapolis: University of Minnesota Press.

Fassin, D. (2011) *Humanitarian Reason*. Berkeley, Los Angeles and London: University of California Press.

Federici, S. (2004) *Caliban and the Witch: Women, the Body and Primitive Accumulation*. Brooklyn NY: Autonomedia.

Foucault, M. (1976a) *The History of Sexuality. The Care of the Self*. Vol. 3. New York: Pantheon (this edition 1986).

Foucault, M. (1976b) *Society Must Be Defended*. New York: Picador (this edition 2003).

Foucault, M. (1999) *Abnormal: lectures at the College de France 1974–1975*. London and New York: Verso (this edition 2003).

Fredrikson, G. M. (1971) *The Black Image in the White Mind: The Debate on Afro-American Character and Destiny, 1817–1914*. New York: Harper and Row.

Friedman, J. B. (1981) *The Monstrous Races in Medieval Art and Thought*. Cambridge, MA, and London, UK: Harvard University Press.

Galli, C. (2010) *Political Spaces and Global War*. Minneapolis: University of Minnesota Press.

Garraio, J. (2019) 'Challenges and backlashes of #MeToo: the case of Germany'. In: Corsi, M., Thissen, L. and Zacchia, G. (eds) *The #MeToo Social Media Effect and its Potentials for Social Change Europe*. Brussels: FEPS – Foundation for European Progressive St, pp. 38–54.

Giubilaro, C. (2018) 'Lo spettacolo del naufragio. Migrazioni, luoghi visuali e politica delle emozioni'. In: Arfini, E. A. G., Deplano, V., Frisina, A., Giuliani, G., Tesfaù, M. G., Perilli, V. et al. (eds) *Visualità & (anti)razzismo*. Padova: Padova University Press, pp. 10–23.

Giuliani, G. (ed.) (2015a) *Il colore della nazione*. Florence-Milan: Le Monnier/Mondadori Education.

Giuliani, G. (2015b) 'Fears of disaster and (post-)human raciologies in European popular culture (2001–2013)'. *Culture Unbound* 7 (3), pp. 363–385. http://doi.org/10.3384/cu.2000.1525.1572363

Giuliani, G. (2016a) *Zombi, alieni e mutanti. Le paure dall'11 settembre ad oggi*. Florence-Milan: Le Monnier /Mondadori Education.

Giuliani, G. (2016b) 'Monstrosity, abjection, and Europe in the War on Terror'. *Capitalism Nature Socialism* 27 (4), pp. 96–114. http://doi.org/10.1080/10455752.2016.1192212

Giuliani, G. (2016c) 'Per una mappatura transnazionale dell'assegnazione del colore tra colonialismo e condizione postcoloniale'. *From the European South* 1, pp. 111–126.

Giuliani, G. (2017) 'The colour(s) of Lampedusa'. In: Proglio, G. and Odasso, L. (eds) *Border Lampedusa. Subjectivity, Visibility and Memory in Stories of Sea and Land*. Cham: Palgrave Macmillan, pp. 67–85. http://doi.org/10.1007/978-3-319-59330-2_5

Giuliani, G. (2019) *Race, Nation and Gender in Modern Italy. Intersectional Representations in Visual Culture*. London: Palgrave Macmillan. http://doi.org/10.1057/978-1-137-50917-8
Giuliani, G. (2020a forthcoming) 'Monstrous beauties: bodies in motion between colonial archives and the migrant and refugee crisis'. In: Tate, S. A. (ed.) *The Palgrave Handbook on Critical Race and Gender*. London: Palgrave.
Giuliani, G. (2020b forthcoming) 'The end of the world as we know it'. In: Giuliani, G. (ed.) 'Cidadania em perigo: crime, fim-do-mundo e biopolítica nas literaturas e no cinema pós-coloniais'. *ecadernos* 32.
Giuliani, G. and Vacchiano, F. (2019) 'Il senso degli italiani (e degli europei) per l'Altrx che vien dal mare. Cinema, bianchezza e Middle passage'. In: de Franceschi, L. and Polato, F. (eds) *Narrazioni postcoloniali della contemporaneità, tra conflitto e convivenza/Postcolonial Narratives of the Present, between Conflict and Coexistence*. Special issue *Imago* 10 (19), pp. 137–152.
Giuliani, G., Garraio, J. and Santos, S. J. (2020a) 'Online social media and the construction of sexual moral panic around migrants in Europe'. *Socioscapes* 1, pp. 161–180.
Giuliani, G., Giametta, C. and Petrovich Njegosh, T. (2020b forthcoming) 'Per un'analisi della memoria delle migrazioni in Europa: discorsi, (auto)rappresentazioni e propaganda'. In: Violi, P. and Salerno, D. (eds) *Migranti, archivi, patrimonio. Memorie pubbliche delle migrazioni*. Bologna: il Mulino.
Graham, E. L. (2002) *Representations of the Post/Human. Monsters, Aliens and Others in Popular Culture*. New Brunswick: Rutgers.
Guillaumin, C. (1972) *L'idéologie raciste. Genèse et langage actuel*. Paris: Mouton & Co.
Hage, G. (2016) 'État de siège: A dying domesticating colonialism?' *American Ethnologist* 43 (1), pp. 38–49. http://doi.org/10.1111/amet.12261
Hall, S., Roberts, B., Clarke, J., Jefferson, T. and Critcher, C. (1978) *Policing the Crisis: Mugging, the State, and Law and Order*. London: Macmillan.
Hall, S. (2007) 'Un dialogo con C.L.R. James (1981)'. *Studi culturali* 2, pp. 237–267.
Hancock, I. (1988) *The Pariah Syndrome: An Account of Gypsy Slavery and Persecution*. Ann Arbor, MI: Karoma Publishers.
Haraway, D. J. (1988) 'Situated knowledges: the science question in feminism and the privilege of partial perspective'. *Feminist Studies* 14 (3), pp. 575–599. http://doi.org/10.2307/3178066
Haraway, D. J. (2015) 'Anthropocene, Capitalocene, Plantationocene, Chthulucene: making kin'. *Environmental Humanities* 6 (1), pp. 159–165. http://doi.org/10.1215/22011919-3615934
Haraway, D. J. (2016) *Staying with the Trouble. Making Kin in the Chthulucene*. Durham: Duke University Press.
Hartman, S. V. (1997) *Scenes of Subjection: Terror, Slavery, and Self-Making in Nineteenth-Century America*. Oxford: Oxford University Press.
Hobson, J. (2005) *Venus in the Dark: Blackness and Beauty in Popular Culture*. London: Routledge. http://doi.org/10.4324/9780203700563
Huysmans, J. (2006) *The Politics of Insecurity: Fear, Migration and Asylum in the EU*. London: Routledge. http://doi.org/10.4324/9780203008690
Iveković, R. (2019) *Politiques de la traduction. Exercices de partage*. Marseille: TERRA-HN éditions Alterego.
Jacomella, G. (2015) 'The silence of migrants. The underrepresentation of migrant voices in the Italian mainstream media'. In: Bond, E., Bonsaver, G. and Faloppa,

F. (eds) *Destination Italy: Representing Migration in Contemporary Media and Narrative*. Bern: Peter Lang, pp. 149–164. http://doi.org/10.3726/978-3-0353-0657-6

James, C. L. R. (1953) *Mariners, Renegades and Castaways. The Story of Herman Melville and the World We Live In*. London and New York: Alison & Busby (this edition 1985).

Jeffreys, S. (2005) *Beauty and Misogyny: Harmful Cultural Practices in the West*. New York and London: Routledge. http://doi.org/10.4324/9780203698563

Kabbani, R. (1986) *Imperial Fictions. Europe's Myths of Orient*. Glasgow: HarperCollins.

Kant, I. (1756) *History and Physiography of the Most Remarkable Cases of the Earthquake which Towards the End of the Year 1755 Shook a Great Part of the Earth*. Hong Kong: Philopschy (this edition 1994).

Kilomba, G. (2018) *Plantation Memories: Episodes of Everyday Racism*. Münster: Verlag.

Koselleck, R. (1979) *Futures Past. On the Semantics of Historical Time*. New York: Columbia University Press (this edition 2004).

Kothari, A., Salleh, A., Escobar, A., Demaria, F. and Acosta, A. (eds) (2019) *Pluriverse. A Post-Development Dictionary*. New Delhi: Tulika Press. https://doi.org/10.1016/j.ecolecon.2017.08.012

Lewis, S. L. and Maslin, M. A. (2015) 'Defining the Anthropocene'. *Nature* 519, pp. 171–180. https://doi.org/10.1038/nature14258

Liebes, T. and Curran, J. (eds) (1998) *Media, Ritual and Identity*. London and New York: Routledge.

Lindqvist, S. (1997) *Exterminate All The Brutes*. New York: The New Press.

Lorde, A. (1988) *A Burst of Light and Other Essays*. New York: Ixia Press (this edition 2017).

Luckhurst, R. J. (2015) *Zombies. A Cultural History*. London: Reaktion Books.

Mai, N. (2014) 'Between embodied cosmopolitism and sexual humanitarianism. The fractal mobilities and subjectivities of migrants working in the sex industry'. In: Anteby-Yemini, L., Baby-Collin, V. and Mazzella, S. (eds) *Borders, Mobilities and Migrations. Perspectives from the Mediterranean, 19th–21st Century*. Brussels, Bern, Berlin, Frankfurt am Main, New York, Oxford and Wien: Peter Lang, pp. 175–192. https://doi.org/10.3726/978-3-0352-6416-6

Mbembe, A. (2001) *On the Postcolony*. Berkeley and Los Angeles: University of California Press.

McClintock, A. (1995) *Imperial Leather: Race, Gender and Sexuality in the Imperial Context*. London and New York: Routledge.

McFadden, P. (2020) Contemporarity: Sufficiency in a Radical African Feminist Life. [Lecture at CES – University of Coimbra] 27 January.

Mehta, U. S. (1999) *Liberalism and Empire: A Study in Nineteenth-Century British Liberal Thought*. Chicago: University of Chicago Press.

Mercer, K. (1986) 'Imagining the black man's sex'. In: Holland, P., Spence, J. and Watney, S. (eds) *Photography/Politics*. Vol 2. London: Comedia, pp. 61–69.

Mezzadra, S. (2001) *Diritto di fuga. Migrazioni, cittadinanza, globalizzazione*. Verona: Ombre corte.

Mezzadra, S. (2004) 'The right to escape'. *ephemera. theory and politics in organization* 4 (3), pp. 267–285.

Mezzadra, S. and Neilson, B. (2013) *Border as Method, or, the Multiplication of Labor*. Durham, NC: Duke University Press.

Mirzoeff, N. (2011) *The Right to Look. A Counterhistory of Visuality*. Durham: Duke University Press.

Mirzoeff, N. (2018) 'It's not the Anthropocene, It's the white supremacy scene, or, the geological color line'. In: Grusin, R. (ed.) *After Extinction*. Minneapolis: University of Minnesota Press.
Mitman, G., Haraway, D. and Tsing, A. L. (2019) 'Donna Haraway and Anna Tsing reflect on the Plantationocene'. *Edge Effects*. https://edgeeffects.net/haraway-tsing-plantationocene/
Moore, J. W. (2015) *Capitalism in the Web of Life: Ecology and the Accumulation of Capital*. New York: Verso.
Moore, J. W. (2016) *Anthropocene or Capitalocene?: Nature, History, and the Crisis of Capitalism*. Oakland: PM.
Moore, J. W. (2017) 'The Capitalocene, part I: on the nature and origins of our ecological crisis'. *The Journal of Peasant Studies* 44 (3), pp. 594–630. https://doi.org/10.1080/03066150.2016.1235036
Moore, J. W. (2018) 'The Capitalocene, part II: accumulation by appropriation and the centrality of unpaid work/energy'. *The Journal of Peasant Studies* 45 (2), pp. 237–279. https://doi.org/10.1080/03066150.2016.1272587
Neocleous, M. (2005) *The Monstrous and the Dead: Burke, Marx, Fascism*. Chicago: University of Chicago Press.
Nuzzo, L. (2013) 'Foucault and the enigma of the monster'. *International Journal of Semiotic Law* 26 (1), pp. 55–72. https://doi.org/10.1007/s11196-012-9275-8
Nuzzo, L. (2004) *Il linguaggio giuridico della conquista. Strategie di controllo nelle Indie spagnole*. Napoli: Jovene.
Pateman, C. (1988) *The Sexual Contract*. Stanford: Stanford University Press.
Pogliano, A. and Solaroli, M. (2012) 'La costruzione visiva dell'immigrazione nella stampa italiana: fotografie giornalistiche e cornici culturali meta-comunicative'. *Studi Culturali* 9 (3), pp. 371–399.
Pogliano, A. (2019) *Media e immigrazione*. Rome: Carocci.
Portinaro, P. P. (1992) 'Materiali per una storicizzazione della coppia amico-nemico'. In: Morani, M., Portinaro, P. P. and Vitale, A. (eds) *Amicus (inimicus) hostis*. Milan: Giuffrè, 1992, pp. 219–310.
Povinelli, E. (2016) *Geontologies. A Requiem to Late Liberalism*. [ebook reader] Durham and London: Duke University Press. https://doi.org/10.1215/9780822373810
Ribeiro, D. (2019) *Lugar de fala*. São Paulo: Pólen Livros. https://doi.org/10.1590/s0104-71832019000200015
Rich, A. (1987) 'Note toward a politics of location'. In: *Blood, Bread and Poetry: selected prose 1979–1985*. London: Virago Press, pp. 210–231.
Said, E. (1978) *Orientalism*. London: Routledge and Kegan Paul.
Said, E. (1993) *Culture and Imperialism*. New York: Vintage Books.
Salerno, D. (2016) 'Risky subjects in time of terror. A semiotic perspective on the security discourse in Europe'. *Versus* 123 (2), pp. 363–384. http://doi.org/10.14649/85508
Santos, R., Roque, S. and Santos, S. J. (2018) 'De-securitising "the South in the North"?' Gendered narratives on the refugee flows in the European mediascape'. *Contexto Internacional* 40 (3), pp. 453–477. https://doi.org/10.1590/s0102-8529.2018400300003
Saucier, P. K. and Woods, T. P. (2014) 'Ex aqua: the Mediterranean basin, Africans on the move and the politics of policing'. *Theoria: The Journal of Social and Political Theory* 141 (December), pp. 55–75. https://doi.org/10.3167/th.2014.6114104

Scarry, E. (1985) *The Body in Pain. The Making and Unmaking of the World*. Oxford: Oxford University Press.

Seymour, N. (2013) *Strange Natures Futurity, Empathy, and the Queer Ecological Imagination*. Urbana: University of Illinois press.

Sharpe, C. (2009) *Monstrous Intimacies: Making the Post-Slavery Subject*. Durham: Duke University Press. https://doi.org/10.1215/9780822391524

Schmitt, C. (1962) *Theory of the Partisan: Intermediate Commentary on the Concept of the Political*. New York: Telos Press Publishing (this edition 2007).

Stengers, I. (2015) *In Catastrophic Times: Resisting the Coming Barbarism*. Lüneburg: Open Humanities Press & Mason Press. https://doi.org/10.14619/016

Stoler, A. L. (2002) '*Colonial Archives* and the arts of governance'. *Archival Science* 2 (1–2), pp. 87–109. https://doi.org/10.1007/BF02435632

Stoler, A. L. (2016) *Duress. Imperial Durabilities in Our Times*. Durham: Duke University Press. https://doi.org/10.1215/9780822373612

Tazzioli, M. (2017a) 'The government of migrant mobs. Temporary divisible multiplicities in border-zones'. *European Journal of Social Theory* 20 (4), pp. 473–490. https://doi.org/10.1177/1368431016658894

Tazzioli, M. (2017b) 'Containment through mobility: migrants' spatial disobediences and the reshaping of control through the hotspot system'. *Journal of Ethnic and Migration Studies* 44 (16), pp. 2764–2779. https://doi.org/10.1080/1369183X.2017.1401514

Tazzioli, M. and Walters, W. (2016) 'The sight of migration: governmentality, visibility and Europe's contested borders'. *Global Society* 30 (3), pp. 445–464. https://doi.org/10.1080/13600826.2016.1173018

Tazzioli, M. and Walters, W. (2019) 'Migration, solidarity and the limits of Europe'. *Global Discourse* 9 (1), pp. 175–190. https://doi.org/10.1332/204378918X15453934506030

Tsing, A. L. (2015) *The Mushroom at the End of the World: On the Possibility of Life in Capitalist Ruins*. Princeton, NJ: Princeton University Press.

Vergès, F. (2017) 'Racial Capitalocene'. In: Johnson, G. T. and Lubin, A. (eds) *Futures of Black Radicalism*. London and New York: Verso, pp. 72–82.

Wekker, G. (2016) *White Innocence. Paradoxes of Colonialism and Race*. Durham: Duke University Press. https://doi.org/10.1215/9780822374565

Wolfe, C. (2005) *Monsters and Philosophy*. London: College Publications.

Yeğenoğlu, M. (1998) *Colonial Fantasies: Towards a Feminist Reading of Orientalism Cambridge Cultural Social Studies*. Cambridge: Cambridge University Press.

Young, R. (1990) *White Mythologies: Writing History and the West*. London and New York: Routledge.

Yusoff, K. (2018) *A Billion Black Anthropocenes or None*. [ebook reader] Minneapolis: Minnesota University of Minnesota Press.

Films

Amiel, Jon. *The Core*. USA, 2003.

Bayona, Juan Antonio. *The Impossible*. USA, 2012.

Blomkamp, Neill. *Elysium*. USA, 2013.

Boyle, Danny. *28 Days Later*. UK and USA, 2002.

Dahan, Yannick and Benjamin Rocher. *The Horde*. France, 2009.

Edwards, Gareth. *Monsters*. UK, 2010.

Emmerich, Roland. *The Day After Tomorrow*. USA, 2004.
Darabont, Frank. *The Mist*. USA, 2007.
Forster, Mark. *World War Z*. USA and Malta, 2013.
Fresnadillo, Juan C. *28 Weeks Later*. UK and Spain, 2007.
Garland, Alex. *Annihilation*. USA, 2018.
Glazer, Jonathan. *Under the Skin*. UK, USA, Switzerland, 2013.
Gwo, Frant. *The Wandering Earth*. China, 2019.
Hall, Stuart. *Race. The Floating Signifier*. UK, 1997.
Joon Ho, Boong. *Snowpiercer*. USA, 2013.
Lawrence, Francis. *I am Legend*. USA, 2007.
McCarthy, Colm. *The Girl with All the Gifts*. UK, 2016.
Meirelles, Fernando. *Blindness*. USA and Brazil, 2008.
Spielberg, Steven. *War of the Worlds*. USA, 2005.
Sputore, Grant. *I Am Mother*. Australia and USA, 2019.
Villeneuve, Denis. *Arrival*. USA, 2016.
Winterbottom, Michael. *Code 46*. UK, 2003.

TV series

Aguilera, Pedro. *3%*. Brazil, 2016– (three seasons).
Darabont, Frank. *The Walking Dead*. USA, 2010– (ten seasons).
Mitchell, Dominic and Jonny Campbell. *In the Flesh*. UK, 2013–15 (two seasons).

Exhibitions

Edward Burtynsky, Jennifer Baichwal and Nicholas de Pencier. Cur. 'Anthropocene'. MAST (Bologna, Italy), 16.05.2019/5.01.2020.

1 *The past devours the present*

Fears of invasion and the repressed memory of colonial violence

1 Introduction

The 'we's' hegemonic imaginary of the human and environmental apocalypse is filled with famished and desperate, aggressive 'floods' of migrants and refugees escaping from poverty, climate change effects and wars (Bettini 2013). Mental pictures of migrant Others as a rapacious invading 'horde' draw on the same material found in apocalyptic narratives about anthropophagic monsters. In fact, the 'body spectacle' reaching its climax with the devouring and dismemberment that accompany the uprising of rabid monsters in a paroxysm of violence conjures up the idea that invaders are tearing apart the (white) body of the imagined community of the 'we' (Williams 1991). Discourses of human and post-human deadly invasions evoke and *reveal* a long tradition of describing barbarians (be they indigenous or enslaved) as cannibals, rationalising invasion and occupation, torture and death in the 'elsewhere' as a means to prevent the godless and uncivilised almost- or post-humans from 'devouring' the God-fearing and reason-driven finer specimens of humankind. Likewise, the hegemonic understanding of migration as a *monstrous invasion* may be read as being the product of the 'we's' 'guilty conscience' about colonial violence and fears that the horrors of colonialism and slavery may strike back. Europe and the West's repressed colonial memory is linked to painful emotional records of past regimes of segregation, sexual violence, forced reproduction and death affecting both victims and perpetrators, which transcend the individual but rather refer to a collective experience mirroring the repressed psychopathological experience of colonial violence shared by the colonised (Fanon 1952).[1]

Repressed memory is the result of historical processes of collective denial and a society's refusal to confront the past. Although long dismissed as an exclusively American phenomenon related to the slave trade or as affecting only those who had a direct experience of colonialism, these memories haunt the present of all peoples, cultures and countries that directly contributed to, benefited from, were created through or influenced by colonialism and slavery. Tracing the genealogy of the anthropophagic monster in literature, in political theory and in popular culture, I will explore the relation between the cannibal in colonial imagination and *monstrous* representations of postcolonial migrants and refugees in contemporary news media.

Focusing on the genealogy of current processes of de-individuation, massification, animalisation and de-subjectivation that make the body in motion 'monstrous', I will trace out connections and disconnections between past and present 'scenes of subjection' and between signifying processes at work in the plantation and at the border. Through a contrapuntal reading of the slave ship crossing the Atlantic and what I see as its reverse *chronotope* – the migrant boat crossing the Mediterranean – I will show that the archive of 'figures of race' that informed relations of subjugation during slavery and colonialism is still active despite a fundamental difference between old and new bio- and necropolitics of the Middle Passage: migration is the result of an autonomous decision. I will explain how the semiotic power of the border is crucial to associating the autonomous journey of the body in motion and its symbol – the boat – with images of catastrophe and death to create *white anxiety* and *moral panic* and legitimise the Mediterranean border regime. Finally, through the concept of 'semiotic border', which I will discuss in more detail in the next section, I will examine the relation between colonial figures of race, migrant crossings and *moral panic*.

As I will argue, besides delegitimising the autonomous journey of racialised migrants, the semiotic border's *monstrification* of migration from the Global South relieves the 'we' of any responsibility for its causes. In the 'we's' hegemonic narrative, the evil effects of the Anthropocene and the violence of the uncivilised are the real culprit in the unrelenting invasion that will make the 'boat sink'. Europe and the West instead are often praised for their benevolence – the immaculate effect of white moral superiority. Because if Europe and the West had no role in causing historical colonial violence, far-flung disasters and morally distant wars (i.e. led by fanatics as opposed to reason-driven), then providing shelter to the desperate might only represent an act of charity on the part of the white saviour. This same benevolence, however, is always accompanied by dire warnings of the catastrophe looming behind more permissive borders: the hungry, uncivilised and ungrateful 'poor' refugees may soon turn into a dangerous internal enemy, envious of the 'we's' well-being, civilisation and culture and set to devour 'us'. In the double-sided discourse surrounding the border, mobility is at once criminalised and regarded with pity – a symptom of colonial duress in our neoliberal present.

Drawing on theoretical articulations and debates in political philosophy and in gender, postcolonial, critical race and whiteness studies, which have provided a rich framework for my previous work on constructions of race and gender in colonial and postcolonial settings, I will read visual news media representations and horror films in light of connections and disconnections between colonial modernity and the neoliberal present. My analysis will develop around the key concepts of 'suppressed memory of colonial violence', 'figures of race', 'human', 'humanness', 'de-humanisation' and 'post-humanness' to *reveal* the main discursive effects of demonising the threat represented by migrants and refugees and describing environmental disasters only in terms of their impact on migration. The outcome is that the West, and Europe in particular, is re-centred in the history of human progress as a mere victim of global phenomena. If the nature and consequences of disasters elsewhere are considered only in terms of their potential negative impact on Europe, it follows that Europe can once again be seen as the bastion of white/Western civilisation besieged by monster invaders, environmental disasters and fanatic wars.

This section is organised as follows. Chapter 2 presents a brief history of Western representations of cannibalism, symbolising evilness and *monstrosity* of the gendered and racialised Other than antiquity through the present – where monsters, and cannibal monsters in particular, are used to demonise individuals and groups as well as non-human entities perceived as threatening. Once considered an inherent trait of indigenous savagery and closely associated with Satan, anthropophagy is a mark of the zombie's 'anonymous fury' in today's popular culture. I will argue for a connection between the widespread revival of cannibalistic mobs and rabid zombies on one hand, and the depiction of migrant landings and terrorist threats in news media and films on the other, inasmuch as the imagery used to represent migrations as deadly invasions signals the white coloniser's repressed memory of his own past cannibalistic invasions into territories that would become the peripheries of his empires.

Chapter 3 connects the repressed memory of slavery to contemporary constructions of whiteness and monstrosity. Specifically, it focuses on the *dispositifs* that in the slave ship, the slave auction and the plantation were used to de-humanise and turn people into slaves and connects them to current practices and discourses il-legalising migrants. After exploring the symbolic materials underpinning the construction of migrants as illegal – that is, the materials feeding representations of the unruly mobility of the racialised monster that is believed to be at the root of terror attacks and economic, social and environmental crises in the receiving countries – I will connect such representations to Europe's repressed memory of colonial and postcolonial violence and capitalist exploitation.

Chapter 4 traces the genealogy of the cannibalistic undead in cinema through a 'political philosophy of the zombie' that outlines its evolution into today's *fast-running* and *devouring* post-human. Finally, it introduces the films and TV series that will be discussed in detail in Chapters 5 and 6: Danny Boyle's *28 Days Later* (2002) and Juan C. Fresnadillo's sequel, *28 Weeks Later* (2007); Yannick Dahan and Benjamin Rocher's *The Horde* (original title *La Horde*) (2009); Dominic Mitchell and Jonny Campbell's BBC drama series *In the Flesh* (2013–14); Francis Lawrence's *I am Legend* (2007), Marc Forster's *World War Z* (2013), and Frank Darabont's ongoing TV series *The Walking Dead* (2010–).

As I will explain in Chapter 7, the fact that all these films and TV series feature *fast-running* and *devouring* post-humans – that is, the zombie's latest evolution into a rabid human looking more like an animal than an undead – was one of the determining factors in my selection, as it allowed me to explore the relationship between the monstrification of the threat, the speed and anthropophagic nature of neoliberal capitalism, and the fast and devouring nature of global wars and climate change. By global wars, I mean an assemblage of discourses and securitisation practices implemented in Europe and the West after 9/11 and, in line with Talal Asad (2007), Judith Butler (2009) and Ann Laura Stoler (2016), a symptom of persisting colonial attitudes. Another determining factor had to do with understanding the popularity of colonial biopolitical *dispositifs* (Stoler 2016) in horror culture – camps, colonies and carceral archipelagos that together with borders and walls are established by the survivors to escape and fight – and the extent to which the colonial imaginary goes unchallenged or, on the contrary, counter-narratives to neo-colonial discourses are provided.

In the conclusion, I will summarize the reasons for employing the *fast-running* and *devouring* monster to investigate the persistence of figures of race in neoliberal reconfigurations of colonial biopolitical *dispositifs* within the frame of the war on terror. I will then expand my reflections on the symbolic materials shaping the imaginary behind European hegemonic fears of migrant monsters and catastrophes.

2 *Hic sunt cannibals*

Monstrosity appears to have signalled a limit, a boundary, or a border since the very beginning of human life. For centuries it was associated with representations of chaos, nature and the *finis mundi*, the end of the world and of what was considered 'right' in Western society (Foucault 1961; Nuzzo 2013, p. 56). In antiquity, monstrosity epitomised the border between those who were considered a direct emanation from God and all the rest – excluded from the *polis* or from the reign of God-like beings (Del Lucchese 2019) – such as women, non-human beings (animate and inanimate nature, including animals), demons and devils.[2] The anthropophagic monster can be traced back to the Middle Ages, when the act of cannibalising the flesh and souls of sinners and infidels was associated with Evil and the anti-Christ. Connecting human-flesh eating to animals and some animals to Satan, cannibals were seen as closer to the reign of Evil than they were to the reign of God in the hierarchy of Creation.

> With six eyes he was weeping and over three chins dripped tears and bloody foam. In each mouth he crushed a sinner with his teeth as with a heckle and thus he kept three of them in pain; to him in front the biting was nothing to the clawing, for sometimes the back was left all stripped of skin.
>
> (Dante, *Inferno*, Canto XXIV, pp. 1304–1321)

Beast-like representations of cannibalistic monsters dominated scholastic iconography and the descriptions of indigenous populations in fifteenth-century travel memoirs by scientists, sailors and pirates who rounded the Cape of Good Hope towards the Pacific. Africa, the Caribbean, Amazonia and the Pacific were thought to be simultaneously Heaven and Hell – a contradiction that was still characteristic of twentieth-century Western culture, from Joseph Conrad's (1899) portrait of the land of cannibals in *Heart of Darkness* to Ruggero Deodato's heavily censored *Cannibal Holocaust* (1981), a proto-snuff film about young American anthropologists exploring the Amazon rainforest to study cannibals.

In early modern colonial discourse, cannibalistic monstrosity was the last evil on the path to creating overseas utopias where the righteous could build the City of God. Monstrosity and utopia became virtually inseparable terms of the narrative sanctioning not only colonialism (Said 1978, 1993) but also enslaved (Allen 1994) and indentured labour (Banivanua-Mar 2005, 2007; Biber 2005; Berglund 2006) as manifestations of God's will.

Allegations of cannibalism and multiple conflicting myths about the indigenous peoples were a matter of debate among scientists, merchants and churchmen who for varying reasons were involved in the New World utopia. Sixteenth-century mythologies depicted Native Americans as either godless humans in need of protection (de las Casas 1565) or 'worse than bloodthirsty beasts' who 'devoran carne humana, andan desnudos [...] entregados a los más vergonzosos delitos de lujuria y sodomia',[3] and for that reason must be annihilated (de Acosta 1577, cited in Langer 2010). In the seventeenth and eighteenth centuries, they were alternately portrayed as expendable, infantile quasi-humans (Harrington 1656), noble savages, indomitable warriors, or bloodthirsty 'primitives' (Lafitau 1724; De Paw 1771).[4] In the seventeenth century, indigenous peoples' primitivism was increasingly interpreted as inability on their part to make the land productive. After the English philosopher John Locke stated this idea as the principle of *terra nullius* in his *Second Treatise on Government* (1690), it became a central issue in the political debate surrounding the conquest of the New World. Because of the indigenous people's inability to cultivate them, Locke claimed, the New World's lands were to be considered *empty* and destined by God for the chosen people to give birth to a new and more righteous society based on production. But *empty lands* (1690, § 37) meant that the indigenous people ought to be considered a *nothing* that could be easily wiped out, save for a few subjected to the rule of Christianity, its political formations, and capitalism (Giuliani 2006). *Hic sunt leones* ('here are lions', usually written on uncharted territories of old maps) and there, *homo homini lupus* ('man is wolf to man'), as Thomas Hobbes (1651) would say (Avramescu 2003): a condition to which even the monstrous Leviathan was preferable. In the coloniser's mind, eliminating the undeserving cannibalistic savages would make 'primitive accumulation' more accessible (Phillipps 1998; Bartolovich 1998) for the benefit of 'the people' – not just lords and landowners as was the case with England's enclosures. A new society would arise, grounded in a social contract among equals (white, male Anglo-Saxon proprietors, faithful to the Crown and its Church). In early seventeenth-century American colonialism, this meant appropriating lands and replacing the indigenous people – who were also considered useless for labour – with a domesticated labour force from other colonised lands or from the swamps of Britain (Creed and Hoorn 2002).

The seventeenth century was also the time when Africa was set to become a trope for cannibalism in European imaginary, and cannibalism legitimised genocide and enslavement for the sake of creating the Americas' Eden – a popular belief that persisted through the nineteenth century, when alleged cannibals from Dahomey were exhibited in European human zoos (Blanchard et al. 2008). In eighteen-century colonial literature, accounts of Captain Cook's exploration of the Pacific and annihilation of many indigenous communities were accompanied by tales of ritual dismemberments and cannibalism, resulting in an increase in Pacific expeditions to free Eden from barbarism (Obeyesekere 1992, 2001; Biber 2005). Interestingly, real or alleged accusations of cannibalism against indigenous communities in the nineteenth-century were often accompanied by evidence of anthropophagy among European crews, as sailors would often resort to cannibalism to survive following a shipwreck (Luckhurst 2015, p. 51).

Figure 1.1 Théodore Géricault, 'Le Radeau de la Méduse' (fragment), oil on canvas, c. 1818–1819.

New figures, or better, new rearticulations of traditional colonial and metropolitan monsters, were forged between the seventeenth and nineteenth centuries. The sexually rapacious and lazy black slave, the Yellow Peril, the rebellious coolie, the lawless and sexually incontinent Arab or Mediterranean immigrant, and feminine or queer monstrosities in Georgian and Victorian England all served the construction of differential regimes of exploitation that were gendered and racialised. From the Renaissance to the Enlightenment, denying the humanity of the enslaved and colonial Other resulted in limitless violence of word, rule and sword. Nevertheless, the actual or imaginary association between the enslaved, the colonised, the criminal and anthropophagy endured across centuries as a colonial legacy to postcolonial times, visible in colonial reports and popular culture.

As we know from literature on the construction of monstrosity in Modern times, monsters were much more complicated figures than a simple projection of evil onto the 'dispensable' Other. Monstrosity indeed has to do with fear of chaos, decay, annihilation, damnation, political crisis and death. It has to do with both terror and horror, with the political and the intimate, the social and the individual (Hurley 1996; Barker et al. 1998; Kirkup et al. 2000; Graham 2002; Avramescu 2003; Berglund 2006; Jáuregui 2008; Christie and Laurc 2011; Jones 2014).

In our times, monsters and cannibalistic monsters have returned as assemblages of images and imaginaries inherited from the history of Western representations of the Other, often through allegories that apparently were only the product of a certain moment in history and a specific context. Since the Second World War and the awareness it brought of the 'anonymous fury' of the Bomb (Calder Williams 2010, p. 80), and especially since the end of the Cold War and the beginning of decolonisation, Europe has been increasingly troubled by fears. An unprecedented sense of vulnerability has spread, caused by social, political, cultural and economic issues connected to Europe's gradual shift from the 'centre' of modern empires and engine of global dynamics to post-modern 'periphery' (Chakrabarty 2000) of new empires (coinciding, in political theory, with the end of Europe's global unilateralism; see Sloterdijk 2005; Haas 2008). Just like its former colonial counterparts, Europe and the West are similarly or potentially affected by global environmental and political phenomena. Mass migrations, terrorism and environmental catastrophe, in particular, often figure in contemporary horror films about contagious diseases, economic downturns, planetary apocalypses, genetic mutations, or weapons of mass destruction, in which zombies, aliens and mutants allegorise some of the characteristics of the ongoing transformations (Brioni 2013).

Among these allegories, I find the 'undead' particularly telling, as it is capable of signifying the present through the image of the past *devouring* it. I read the zombie as a repressed (colonial) memory resurfacing – that is the return, albeit in a role-reversal of sorts, of the memory of invasion, expropriation, genocide, slavery, disaster and death that Europe and the West have long been unable or unwilling to deal with. It also serves as an allegory for Europe's guilty conscience re-emerging alongside the rising undead – as though Fanon's black consciousness (1952) had escaped from colonial and post-slavery selective memory to reincarnate in the form of mortal enemies. In my view, the undead in popular culture, and particularly the *fast-running* and *devouring* rabid post-humans that materialised after 9/11, signal a looming catastrophe, as if the memory of future apocalypses was an allegory for the 'horrors of history' (Lowenstein 2005, p. 1). Or, indeed, as if the infected *devouring* rabid monster and the *fast-running* undead personified unbounded retaliation, lifting the lid on the unacknowledged horrors endured by the exploited victims of the 'civilised' world.

Therefore, what gets exteriorised or alienated here, using Jacques Lacan's words describing *jouissance* (1994, p. 281), is not the colonial/enslaved Others, with their own autonomous experience, present lives and past memories. Rather, it is the collective *intimate* memory of, and sense of guilt for, the violence the 'we' inflicted on them that assumes the form of *fast-running* and *devouring* rabid monsters. The term 'we', here and elsewhere in my analysis, is meant to indicate a historical community that encompasses Europe and the West and benefits today from the outcome (material *and* symbolic) of epistemic and physical violence. It also connotes an imagined community that consolidates itself and mobilises in response to fears and perceived threats, as in the case of anticolonial wars in the past and acts of terror in the present. This 'we' exteriorises (as well as denying) past acts of violence against the Other, projecting them onto a monster figure perceived as determined to hurt 'us' – as if the monster did not belong to the history of the 'we' but was coming from 'elsewhere', and this 'elsewhere' was not a space and a history of Otherness constructed by the 'we'. As such, the monster's alleged blind, fanatic, animalised violence justifies the legitimacy of the we's' fences and military ferocity. Žižek's (2008) interpretation of Lacan's *jouissance* can help us read this 'exteriorisation' through the lens of the horror genre as transgression and as an urge to bring death to the national (and internal) enemy, as well as understanding why killing zombies is fun for audiences. First, they are portrayed as a repulsive, blindly violent version of the 'we'. Second, they are amongst us, and their defeat coincides with, and is legitimised by, the preservation of the group, that is, the 'we', the real humans. The will to kill post-human beings reinforces the audience's status as worthy real humans.

British sci-fi writer Herbert G. Wells (1898) shared this interpretation of horror films as representations of repressed memories of colonial horrors and disquieting fantasies of counter-invasion. As he wrote in Book One of *War of the Worlds*:

> before we judge of them [the Martians invading the earth] too harshly we must remember what ruthless and utter destruction our own species has wrought, not only upon animals, such as the vanished bison and

> the dodo, but upon its inferior races. The Tasmanians, in spite of their human likeness, were entirely swept out of existence in a war of extermination waged by European immigrants, in the space of fifty years. Are we such apostles of mercy as to complain if the Martians warred in the same spirit?
>
> (Wells 1898, p. 7)

Building on Agamben's insight, Ghassan Hage (2016) has called this fantasy, which he believes was once specific to settler colonialism but is now widespread in Europe and the West as a result of the migrant crisis, 'état de siège'. This sense of living in a state of siege is intimately related to the fear of reverse colonialism underlying Modern discourses on the need to colonise barbarians lest they invade and destroy the civilised world. The sense of besiegement and *white anxiety*, however, were not limited to settler colonialism but also characterised exploitation colonialism (Stoler 2016, p. 202) and the colonial imaginary ridden with fantasies of racial revenge (Luckhurst 2015, p. 63). I will explore this specific reading of past and present fears of counter-invasion in Section 2, focusing on the figure of the alien. While the figure of the undead is particularly meaningful for analysing monstrous representations of migration and the post-9/11 terrorism threat, that of the alien can help read present fears that the geographical and cultural distance between the former coloniser and the colonised may not be maintained.

The presence of repressed memories[5] – both individual and collective – is suggested by the very limited public debate on Europe's colonial past and collusion in slavery and by refusal at the institutional and societal level to read current power relations and forms of discrimination, including those underpinning the border regime, as legacies of colonialism and slavery (Blanchard et al. 2005; Puwar 2004; Jensen and Loftsdóttir 2012; Wekker 2016; Stoler 2016; Ponzanesi and Colpani 2016; Giuliani 2019; Jensen 2018; Sousa Santos 2018). Nevertheless, the same 'figures of race' that legitimised the violence of colonialism and slavery populate today the increasingly racist transnational imaginary that sustains old and new walls and camps, forms of spatial segregation, detention, deportation and differential inclusion of racialised migrants and citizens.

My research on visual representations of migrant landings and border crossings in the local and international press as well as on online platforms (newspapers, Facebook and blogs) also points to such a reading. Fear of the enemy penetrating the space of the 'we' or already settled – the racialised migrant, the internal Other, or the terrorist threat that former U.K. prime minister David Cameron compared to *cancer* in 2010[6] – often represented as *devouring* Europe and the West from within and without, is reproduced through written and visual language creating a semiotic landscape grounded in ideas of catastrophe, symbolic cannibalism, contagious diseases and death. Contemporary monsters embody the speed and anthropophagic nature of neoliberal capitalism and the fast and devouring nature of global wars and climate change impacts. In this picture, monstrosity and catastrophe are integral to each other from the viewpoint of the 'we'. The application of critical discourse analysis to texts about the undead is meant to *reveal* the positionality from which the crisis/catastrophe is imagined and the relationship between the civilised and the uncivilised, as allegorised by the living and the undead.

In fact, not only do *post-human, devouring* hordes suggest what symbolic materials feed *moral panic*, but they also allow constructing by contrast, or at least imagining, the kind of survivor community the 'we' wants or fears identifying with. This community is a representation of the society the 'we' lives in, and indeed the same kind of society constructing the Other as a monster and their mobility towards, and penetration of, Europe and the West as deadly. The question here is, what Europe (and broader West) is that which a horde of monsters puts in danger? What values, self-perceptions, power relations, gender norms and memory is this Europe seemingly based on?

By analysing the ways and contexts in which the devouring monster is created and signified, I will investigate biopolitical devices and practices of containment of the 'monster in mobility', including border security, necropolitics and necropolitics at sea. Military security at the frontier (Huysmans 2006) and necropolitics (Mbembe 2003) are well represented in all zombie films, and particularly in *28 Days Later*, *28 Weeks Later*, *World War Z* (hereinafter *WWZ*) and the TV series *The Walking Dead*. I will also try to understand how subjects in mobility today elude the practices of containment used in the past to confine the colonial and postcolonial Other – not only the camps and colonies analysed by Agamben (2003) and Stoler (2016) but also borders – and how elusion and trespassing transform them into monsters. The 'logic of encampment' (Gilroy 2002) is particularly visible in *28 Weeks Later* and *WWZ*. Films on alien invasions like *District 9* and *Monsters* (Section 2) and films on environmental disasters like *Blindness* (Section 3) provide a very effective allegory for this biopolitical *dispositif*. For that reason, my reflections in Section 1 will be in a circular relation with the other sections of the book, in the attempt of 'blasting open the continuum of history and allegoric correspondences', paraphrasing Adam Lowenstein's adaptation of Walter Benjamin's words (Benjamin 1940; Lowenstein 2005, p. 12).

Through the figure of the *fast-running* and *devouring* post-human, I will investigate representations in popular culture of the practices of bouncing migrants around within the Mediterranean hotspot and detention systems, described by Martina Tazzioli (2017b) and others, as well as surplus-value extraction through their mobilisation between those hotspots and Europe's borders. The first aspect is particularly visible in the more traditional *The Walking Dead*, where the undead are driven from place to place to trigger reconfigurations of geopolitical relations among survivor communities.

Surplus-value extraction from migrants' constrained and/or forced mobility, on the other hand, is nearly unrepresented in popular monster-invasion films. I believe that this lack of thematisation is due, in fact, to a temporary unrepresentability. If it is true that the figure of the zombie as a running/walking dead who cannot be enslaved nor paid a wage is an allegory for this new form of value extraction focused on containment, channelling and obstruction of free movement, then the real purpose of the logistics of migration and migrant containment and displacement is still not adequately represented in most horror films. This is also true of the alien figure, whose allegoric mobility towards Europe is either brought to a halt or contained in camps but never channelled to extract value. In art, as Federica Mazzara (2018) has shown with regard to Mediterranean crossings and shipwrecks, the focus is more on necropolitics/rescuing and on counter-narratives about the Middle Passage, confirming the missing reflection by popular and critical culture on the border system's exploitation of channelled mobilities.

3 *Living dead* and *devouring monsters*

Before proceeding further with my postcolonial and cultural examination of the links between monstrosity and fears of the *past devouring the present*, I need to return to the Modern history of horrors. In this chapter, I will focus on the relationship between the figure of the enslaved in European and North American history and the figures of the undead and the rabid monster, and what this relationship entails in the way of migrant representations as invaders *from the past*.

The plantation lies at the root of the figure of the living dead – modified and re-signified in popular imaginary, literature, cinema and television. But my interest reaches beyond the cultural roots of the zombie to encompass the memory of the absolute violence inherent in treating another human as property, and the escape and rising of the enslaved violating the spatial confinement that constituted them as such and thus becoming monsters. There is more to enslavement than the reduction of human beings to objects on the basis of their alleged non-humanness: the constant reproduction of such non-humanness through daily humiliation and death to hide their proximity to the master and construct the latter as superior. Slavery produces slaves. It is a specific conception of humanness that has produced the not- or less- or post-human since the epoch called the Plantationocene (Yusoff 2018; Davis et al. 2019; Davis and Todd 2017). Hartman (1997) and Sharpe (2009) have brilliantly exposed the vast range of practices aimed at dehumanising the enslaved and taming them into submission: from forcing them to jump naked on the auction block to dictating that slave coffles must sing while walking lest someone cry after being sold, from minstrel shows, rape and forced copulation with other slaves to whipping, lynching and dismembering. All these practices were used to 'make the monster' and, at the same time, they produced white humanness, condemning whites to a vicious circle of violence on which their superiority was ultimately based. These reiterated practices turned the enslaved into socially dead persons, or 'living dead', deprived of emotions and autonomy. Such emotional expropriation was simultaneously legitimised and denied on the grounds that slaves were not human – that is, they were deemed to have reduced emotional activity, immunity to pain, innate seductiveness (e.g. being complicit in their own rape, an aspect of their alleged bestiality) and *non-autonomy* (Scarry 1985, p. 33; Hartman 1997, p. 51).

So long as the enslaved were closely monitored and forcibly confined, the *monstrosity* of the system that made them monsters was concealed in this order of discourse – only to become apparent when they would escape the yoke of slavery, commit murder or kill themselves. As Toni Morrison reminds us in her 1987 novel *Beloved*, Sethe, the protagonist, kills her daughter to spare her the *monstrous* fate she suffered herself. The slave master and white society attribute her crime to her own madness and monstrosity – because her decision to escape is thought to return her to her natural state of crazy monster when, in fact, it is in the very act of subversion that she has regained her humanness and proved the *monstrosity* of slavery.

Similarly, tales of slave mutinies and revolts – such as those described by Cyril James (1938) and Theodore Allen (1994) – are represented as caused by and demonstrating the monstrous nature of the enslaved.

> These coarse men [the blacks] are incapable of knowing liberty and enjoying it with wisdom, and the imprudent law which would destroy their prejudices would be for them and for us a decree of death.
>
> (James 1938, p. 114)

> Many evil disposed servants in these late times of horrid rebellion taking advantage of the looseness of the times, did depart from their service, and followed the rebels in rebellion.
>
> (Allen 1994, vol. 2, p. 619)

In recent times, political philosopher Filippo del Lucchese (2014) discussed the 2010 Rosarno riots by African fruit pickers in the Italian southern region of Calabria, drawing a clear parallel between the enslaved and the overexploited migrant workers. Interpreting both conditions through the political meaning of the riot in the sense of the Spinozian *vindicare*, he wrote that 'whereas the slave himself or herself is the property of the master, only the labour of the employee becomes the property of the employer, no matter what the salary is' (p. 550). I would add that another link between slave and migrant revolts is that both were and are characterised in the white mind as quintessential acts of subversive *monstrosity* (Giuliani 2020). Yann Moulier-Boutang's (1998) and Sandro Mezzadra's (2001, 2004, 2015; Mezzadra and Rahola 2015; Mezzadra and Neilson 2013) work on political genealogies of colonial and metropolitan proletarian revolts can shed light on this link. Contributing a reflection on race and whiteness to their Marxist perspective, I may add that the white male gaze and *white anxiety* for the loss of privileges have always seen black revolts (and mobility across borders meant to identify and contain race and subalternity) as acts of horrific subversion of the white and patriarchal 'order of things'.

Building on the idea that the slave ship is the chronotope of modernity (Gilroy 1993, p. 17), through which a differential ontology is established within the distinction between those who can be reduced to property for surplus value extraction and those who cannot, I would say that the slave ship, the slave auction and the plantation are the semiotic *dispositifs* that constructed race as inherent monstrosity, the violence and unmasterability of which need to be restrained and revert to enslaved docility. We can then draw a parallel between the slave ship and the migrant boat crossing the Mediterranean to Lampedusa and Europe's southern shores. The latter appears as a sort of postmodern reverse chronotope. The *coerced* journey on the slave ship and life in the plantation, described by Douglass (1881, 1855), hooks (1981), Hartman (1997), Rediker (2007) and Sharpe (2009), define the ontology of the enslaved, building by contrast the master's *humanness* as *autonomy*. The migrants' autonomous journey, as well as their living condition in self-built camps, squats and abandoned hangars, like the 'Jungle' in Calais or similar camps in Ventimiglia, Rome, Paris, Idomeni and Patras (Agier 2018), signals the impossibility of that dichotomy.

> autonomy of migration is an insurgent configuration of ordinary experiences of mobility emerging against the regime of border control. At its core is the sharing of knowledge and infrastructures of connectivity, affective cooperation, mutual support and care among people on the move.
>
> (Papadopoulos and Tsianos 2013, pp. 178–196)

Unlike the slave ship, the boat does not signify the constraint that secures the privilege to directly or indirectly benefit from ontological and spatial distance from, and physical subjugation of, the enslaved but rather the precarity of that same boundary/border (see Tazzioli 2015, 2017a; Tazzioli and Walters 2016). In Western and European slavery and post-slavery systems, the coerced nature of the journey that sees the enslaved tied down in the hull of the ship, then standing in chains at the auction block, and, finally, permanently bound to the plantation, has established that dichotomy. But with migrations to Europe, free subjects moving across continents and seas, facing countless obstacles to their movement (constituting what has come to be called the 'border regime', which nonetheless produces violence and death; Jones 2016) and landing autonomously on European shores, bring an end to the illusory idea that Europe is geographically and historically isolated from the horrors of history. The boat is believed to *free* the monster, and as such it triggers *white anxiety* and *moral panic*.

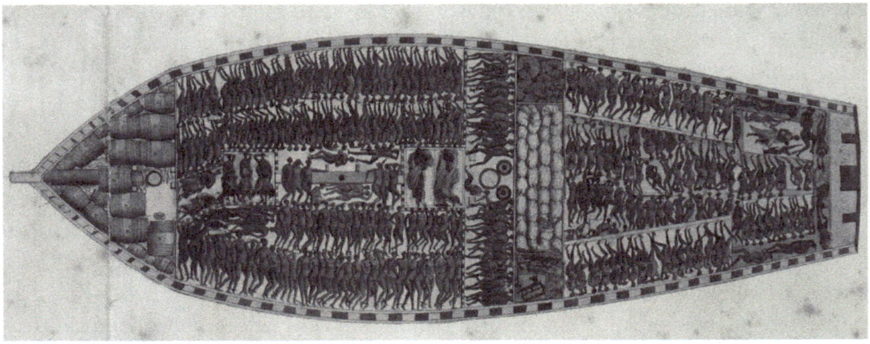

Figure 1.2 Bertrand Guillet, 'La Marie-Séraphique' (fragment), aquarelle on paper, 1772–1773.

Figure 1.3 Massimo Sestini, 'Migrants off the coasts of Lampedusa', 7 June 2014.

Because of this substantial historical and social difference, parallels between the Atlantic and the Mediterranean *Middle Passages* based on the similar condition of slaves and migrants, and humanitarian narratives based on this assumption, are historically and politically problematic (Saucier and Woods 2014, pp. 64–5; Sharpe 2016, pp. 25–62).[7] When talking about a Black Mediterranean, Saucier and Woods, among others, have the history of colonial Europe on their mind, from an Atlantic perspective. The adjective 'black' in Black Mediterranean refers thus to a historical legacy (that of the Black Atlantic of slave trade and slavery) or to something through which, by contrast, Europe constructed itself as white in the colonial past and continues to construct itself as such in the postcolonial present. It does not refer to the people who cross the Mediterranean today, who are not solely black. What Saucier and Woods impressively highlight is precisely my point, that the policing of the Mediterranean is racialised and colonial in nature: the violence of the border against autonomous migrants connects hegemonically disconnected stories of colonial pasts and postcolonial presents inasmuch as this violence both assumes race as a factor in its operation and constantly reproduces it. Nevertheless, I believe that the juxtaposition of the two Middle Passages in 'connecting disconnected' stories of slavery and bounded freedom does not acknowledge important contextual differences at the root of European fears of invasion and *moral panic*.

In fact, from the viewpoint of migrants, the 'antiblackness' of the Mediterranean border regime that Saucier and Woods refer to risks obliterating the constitutive difference between the slave trade and their autonomous project of mobility. While smugglers and anyone profiting from illegalised migration make mobility a constrained and violent process, the choice to flee is made by autonomous subjects who take their chances regardless of the hardships they will face (Nyers 2015). The criminalisation of migrants' autonomy through the border regime, besides being symptomatic of a persisting colonial duress in its necropolitics, is rather the result of *postcolonial and neoliberal* rearticulations of colonial *dispositifs* that aim at denying migrants' full subjectivity and equality with European citizens at a time and place in history where old patterns of colonial domination cannot be publicly acknowledged. Seemingly silent and undetectable forms of oppression are meant to conceal the violence of criminalisation behind a veil of 'compassion, care, empathy and love' (Danewid 2017, p. 1677):

> by divorcing the ongoing Mediterranean crisis from Europe's long history of empire and racial violence, these left-liberal interventions ultimately turn questions of accountability, guilt, restitution, repentance, and structural reform into matters of empathy, generosity, and hospitality. The result is a politics of pity rather than justice, to borrow the words of Hannah Arendt, and a consequent recasting of the relationship between the oppressor and the oppressed as one between the lucky and the unlucky.
>
> (Danewid 2017, p. 1681)

Building on Saidiya V. Hartman's critical analysis of 'empathy' in the context of slavery in the United States, I will discuss how compassion and empathy in the humanitarian gaze ultimately hide not only the monster's subjectivity but also the border's responsibility in producing their suffering. The parallels I draw between the enslaved, the 'channelled' migrant and the *fast-running* and *devouring* horde are based on the idea that the image of the mob of suffering, infected zombies as dangerous invaders comprises an assemblage of meanings and historical fragments from the colonial and slavery archives (Cunningham and Warwick 2018, p. 178). The suffering of the *fast-running* and *devouring* horde – just like the suffering of the enslaved and the colonised and all other figures of race – is a consequence of practices that were designed to reduce their humanness, criminalise them and deny them the right to have the same feelings as (white and Western) humans.

The figure of the zombie is very versatile in its capacity to recall past catastrophes and present fears (Bishop 2010, p. 26). Some have argued for a link between practices of zombie containment and extermination, the memory of the Holocaust and the states of exception during the Second World War (Baishya 2011). Based on scholarly research that has traced the origins of the Holocaust and camp mentality to early colonial Modernity (Friedlander 1997; Stoler 2016), I argue that fears of the present and apocalyptic fears also need to be framed in the historical context of postcolonial rearticulations of colonial biopolitics, because the so-called war on terror and practices of securitisation implemented in Europe and the West after 9/11 are both symptomatic of colonial durability inasmuch as they articulate and essentialise the dichotomy between the 'we' and 'the Other'. The imaginary of the war on terror is accompanied by a renewed fear of chemical and biological warfare, which may modify not only a target subject's body but also their ontology, turning them into a *rightful* target of weapons of mass destruction. Such weapons were first used against the colonised, reinforcing the symbolic difference between the lives (and spaces) worthy of 'respect' according to international laws and those that could be freely devastated. Rabid zombies (themselves the result of biological experiments) attacking the former great powers and colonial empires represent a sort of reverse allegory for the modern and colonial signification of worthy and dispensable lives (Butler 2009; Asad 2007; Asad et al. 2009).

In the next chapter, I will provide a historical overview of zombies in film and discuss some of the more relevant in reproducing the imaginary of deadly invasions and repressed memory of violence.

4 The rise of the undead

The word zombie can be traced back to *nzumbe*, meaning 'god' in Kikongo – a Bantu language spoken in Congo and surrounding areas – and indicates a person who has been brought back to life or to a liminal state between life and death through voodoo rites. It is believed to have come to the Americas with West African enslaved people on the Middle Passage to the West Indies – Haiti first, and then other French and British Caribbean islands.

Zombies first appeared on U.S. screens in the 1930s, though at first there was no mention of cannibalism – absent in films such as *White Zombie* (1932), starring Bela Lugosi as a voodoo witch doctor; *Ouanga* (1936), set on a plantation in Haiti; *King of the Zombies* (1941), a spy story featuring Nazis and black zombies; and the famous *I Walked with a Zombie* (1943), where the undead are the manifestation of a community mourning the horror of slavery. A sense of guilt for the pain suffered by black people in the Caribbean and fear of retaliation from their descendants are palpable in all these films, most of which are set in Haiti, the first black-led republic in history:

> This revolt began in 1791, started by a group of 'Black Jacobins' inspired by the rhetoric of universal liberty promised by the French Revolution; it had to defeat French, Spanish and English troops and a last Napoleonic attempt at reinvading the island before winning final independence.
>
> (Luckhurst 2015, p. 33)

The western obsession with voodooism – part attraction, part repulsion – is seemingly related to the repressed memory of the Haitian revolution and to anti-colonial movements and struggles that started emerging in the Caribbean in the 1930s–1940s. Among the leading figures critical of colonialism were Aimé Césaire and Franz Fanon in Martinique, and Marcus Garvey (later a major influence on the Harlem Renaissance) in Jamaica, who used the metaphor of the zombie in their condemnation of oppression (on Western perceptions of voodooism and its association with vampirism and zombiism, see Kee 2011, pp. 11–13; Luckhurst 2015, pp. 17–97). Nevertheless, it was only with U.S. director George Romero's first film, *The Night of the Living Dead* (1968), that the idea of the dead devouring the living as a payback for pains they suffered in life came to be considered a post-human allegory for a *rightful* social fight.

The figure of the zombie thus comprises a number of interrelated transnational cultural legacies that dominated, resisted, interpreted and appropriated each other. Nevertheless,

> If the zombie emerges in the slippage between cultures, all the same this is not a preface to celebrating the zombie as some kind of sliding signifier that can mean anything we want it to mean. Wherever it comes to stop, the zombie is still branded by the murderous history of slavery and colonial dispossession that underpins its origins. It remains connected to the meaning of Haiti and the islands of the Antilles to the modern world, and the systematic violence, expropriated labour, rebellion and revolution in those areas, however far it travels. What is complex about the figure is often the way this atrocious undertow is at once avowed and disavowed as the zombie stumbles through very different cultures.
>
> (Luckhurst 2015, p. 15)

Therefore, contrary to what Kyle W. Bishop (2010) argued, the zombie is not an 'authentic American product'. Like all cultural constructs, it is to some extent inauthentic and surely transatlantic in its very origin. As such, the zombie finds itself at the intersection of several symbolisms, among which the Protestant Gothic plays an important role (Luckhurst 2012). It draws many of its features from a constellation of Others in sci-fi and fantasy identified over time as enemies or monsters (from vampires to werewolves to robots and aliens; see Browning and Picart 2009), mirrors of our own evil (on the 'more human than human' trope, see Sobchack 2000, p. 138), friendly outsiders like Steven Spielberg's *E.T.* (1982), or revolutionaries fighting for social change (Grebowicz 2007). There is also a peculiarity to the zombie that is found in the binding together of different dimensions: the familiar, because the dead are often friends and family members; the uncanny (i.e. the condition of the living dead); colonial monstrosity and the memory of it. Other figures in horror film provide different representations of Modern Otherness and fears: from vampires and aliens as migrants (though vampires often symbolise 'the horrors of modernity and decadence in a conservative value system', Browning and Picar 2009, p. xvi; Teti 1994); to cyborgs and clones representing human mutations and metamorphoses; to werewolves epitomising the nature/culture, civilised/uncivilised dualisms in white Western *petit bourgeois* culture (Kirkup et al. 2000; Christie and Lauro 2011; Giuliani 2016).

Starting in the 1930s, zombie films featured a succession of mindless resuscitated corpses, until director George Romero revolutionised the canon with *Night of the Living Dead* (1968). In his 38-year-long career, he narrated the return of America's wretched and outcasts as cannibal zombies, exposing the flaws of modernity and consumerist culture: social inequality, individualism and marginalisation. His zombies gradually turned into what political philosophers Antonio Negri and Michael Hardt (2000) would call a 'revolutionary multitude' poised between indistinctiveness and self-differentiation or, in Marx's words, modern 'individuation' – that is, having an acknowledged juridical persona, property rights, inviolable bodies, and subjectivity. This 'rising multitude', in Romero's films as in Negri and Hardt's political theory, embodies what Roger Luckhurst has identified as 'the revolutionary spirit [that] come[s] back and bite[s] you' (2015, p. 157).

Romero's multitude reaches its highest level of self-differentiation and mutual empathy in one of his last works, *The Land of the Dead* (2006). An African American undead called Big Daddy, who was a victim of 'white liberators' (an allegory for the lynching mob) in the *Night of the Living Dead*, leads the multitude in this film, in which zombies grieve for those who were killed (Towlson 2012). The multitude has nothing to lose here and does not need to negotiate: its presence is enough to subvert the laws of the living and after attacking their own city, the zombies walk away together as a newly formed community. In the final scene, the leader of the humans stops another survivor from shooting Big Daddy, saying that 'they are just looking for a place to go, same with us'. Romero's zombies fight for a new society where the relationship with nature is one of equals and communion, support and equality are shared values. There is no production and hence no capitalist accumulation in their vision, no social divisions nor hierarchies, gender relations, sexual reproduction, or time-space boundaries.

In a way, Romero's multitude is diametrically different from the monstrous mobs populating European and Western cities at the end of the nineteenth century and described by conservative philosophers as amorphous, instinctual, beastly, and easily influenced. On the other hand, his zombies (who are obviously *undocumented*) are not fighting progress and individuality as such. What they are challenging is the essence of a country steeped in Modern ideas of subjectivity and national belonging, offering in the end a new, post-human conception of society.

The zombie apocalypse challenges the idea that a 'better' human society is possible in the aftermath of a crisis. In fact, the structural nature of the post-apocalyptic condition progressively erases all traces left by human societies. A new space-time dimension unfolds, in which everything is enveloped in a sense of permanence, leading audiences to think that a society that welcomes the living and the undead is the only chance for survival. What sets apart Romero's films from more recent features inspired by the global war on terror is the way he addressed the contradictions and growing inequality in U.S. capitalism. While Romero questioned state, military and patriarchal authority (Towlson 2012), especially in *Night of the Living Dead* (1968) and *Dawn of the Dead* (1978), zombies today are more frequently deployed in film to signal the enduring need to eliminate monstrosity – a consequence and symbol of the misdeeds of the living – to start anew. Yet the forging of a new elite society based on the survival of the fittest requires annihilating the enemy from the netherworld, the new Caliban 'who has a voice but no language' (Ashcroft 2008; Cunningham and Warwick 2018, pp. 177–178), who roars like an 'animal' and cannot write, as well as being naturally inferior – like the barbarians to the Greeks or the indigenous populations to Christopher Columbus and the conquistadors. One such example is *REC* (2007), a Spanish found-footage film in which an epidemic is attributed to Satan, whose evil plan is to bring the dead back to life. The infernal threat, however, becomes the opportunity for a new Leviathan. Another fine example is *WWZ*, the first blockbuster zombie film, which I will analyse in depth later in this section. Despite many similarities with the poetics of Romero, the main argument in the film is that the 'state of war' will lead to restoring an idea of progress that does not include post-humans, seen as the ultimate enemy. Unlike Romero's films, neither *REC* nor *WWZ* seem to consider post-Fordist capitalism, military power, international politics, and conflicts as greater threats to humanity. The post-apocalyptic society they envision plays into a new eugenicist idea of survival of the fittest where the leader is always white and heteronormative, and either drawn from a bourgeois elite or from the virile working class. The leader is ontologically white, especially if a member of the working class, and fully complies with the normative idea of American hero. Exceptions are made for members of the multicultural upper class and the upper echelons of the military (e.g. in *I am Legend*) who can be 'whitened' blacks – and yet are still portrayed as stereotypically African American (Aprilia and Soelistyarini 2016).

Both *WWZ* and *I am Legend* feature quasi-zombie mutants made rabid by a virus. The mutation that has characterised the figure of the zombie since 9/11 has already occurred before the opening scenes. In the next chapters, I will discuss the symbolic implications of that mutation from living dead to *fast-running* and *devouring* post-human, and the reasons why the 'imagined catastrophe' embodied by contemporary zombies is an allegory for the more likely apocalypses of the present. I will also examine some of the more recent productions in this genre that provide a deeper critique of the ontology and normativity of a postcolonial global system of power which is multi-layered and multi-located (for an overview of 'imagined catastrophes' in sci-fi, horror and Western culture at large, see Seed 2000).

In *The Walking Dead*, the Romero-style stumbling dead apparently turned into zombies after their death as a result of an epidemic. The main theme is the negative memory of past colonial expansion and internal colonisation in the United States, but the film posits the Far West as an *impossible* destination in the present. The survivors are trapped in Georgia, where the whole series is based, struggling to fight zombie attacks on their safe colonies and internal competition for necessities and other resources.

In none of these films – in which survival is alternately dependent on a serum that would break the mutants' fever (*I am Legend*), a vaccine to protect soldiers from zombie attacks (*WWZ*) or a Hobbesian war of all against all (*The Walking Dead*) – is there any empathy whatsoever for the monster.

Is there any chance of – or need for – reconciliation with the absolute (post-human and anthropophagic) Other? As I have argued elsewhere (2016), the BBC series *In the Flesh* offers a shift in a new direction, from the Foucauldian 'sovereign power's right to kill' to a state's power to 'let live' and then 'rule over the bios'. Here, post-zombies are individualised and subjectified (Foucault 1976). To a greater extent than in Romero's films, their condition is tamed with medication. They recover their memories and are re-inserted into their former families. Borrowing Foucault's own words, the partially deceased become a 'governmental object' – they need to live to let the living reconcile themselves with the trauma of the rising, and, more important, with the social violence that caused the death of the loved ones who have now returned. At the same time, they become *bare (post-)life* (Agamben 1995) – a very particular *bios* (that is, living being), if we follow Italian philosopher Roberto Esposito's articulation (2008) – and need medication, injected daily into their spine by doctors and family members, to help re-code them as fine biological and social specimens of humanity. Government-mandated intervention lets post-rabid quasi-humans morph from overwhelming threat to living memory of a permanent mutation (from human to post-human) affecting all contemporary societies.

Without engaging in an in-depth analysis of *In the Flesh*, I will use the TV series as a case study for a contrapuntal reading to understand when and how the post-human is meant to be 'accepted' and 'included', regaining the 'right to speak' they lost in the anthropophagic stage. Through an analysis of the semiotic power of inclusion that transforms the cannibal into an apparently docile (but never fully reliable) member of the community, I will explore the humanitarian narrative against border restrictions that is deployed in the frame of the so-called 'migrant crisis'. I will highlight, in line with De Certeau (1975), the representation of language in Western culture indicating the level of humanness of the subaltern and the zombie *alike*, and the representation of its absence in visual representations of migrant landings in Lampedusa (Casid 2018). I will discuss the contrapuntal relation between empathy towards the suffering of the enslaved, critically examined by Hartman (1997), compassion for the post-humans exposed to death (*28 Days Later*) or recovered to social life (*In the Flesh*), and compassion and empathy in the humanitarian discourse.

5 Post-colonial apocalypses

The French film *The Horde* (2009), co-directed by Yannick Dahan and Benjamin Rocher, represents perfectly what Roger Luckhurst (2015) argued in *Zombies. A Cultural History*: 'in places of poverty, border insecurity and violence, the old anthropology of the *zombie* is still very much alive, dead, and kicking' (p. 169), and also that 'the zombie is [...] animalistic, charged, and feral, and instead of being predictable, it's now inscrutable and volatile' (Dendle 2012, p. 7). The film opens with the funeral of a police officer where a group of desperate mourners, French and North Africans colleagues of the deceased, set out to avenge his murder. Aurore, a white police officer who is expecting a child from the deceased, joins the punitive mission. Their goal is to exterminate the 'Nigerians' who killed their partner. At sunset, they arrive at a dilapidated building in the Parisian banlieue from where a criminal gang comprising two Nigerians, a Pole, a Romani, and two white men acting as lookouts, run their operations. Here, the city is 'provincialised' (Chakrabarty 2000) and the suburbs, normally ignored by the city, are placed in the foreground. Later, on the terrace of the building, the group realise that Paris is under attack by a crowd of *fast-running* and *devouring* rabid monsters (the origin of the epidemic that turns humans into cannibalistic animals is not explained) and see the city in flames from afar.

This impression is confirmed when the group tries to leave the building: the good (the police) and the bad (the drug dealers) team up against the zombies, in line with a certain tradition of violent, urban films such as U.S. cult director John Carpenter's *District 13* (1976). On their way out, they meet an Indochina war veteran, a repulsive old man whose colonialist gestures, expressions, and behaviours recall past constructions designed to dominate or exterminate the 'barbarians' (Lindquist 1997). The veteran proceeds to kill zombies, or 'dirty gooks' as he calls them, with a machine gun and after killing a good-looking one in the hallway, he suggests that his new comrades 'do her', leaving them puzzled and disgusted. In fleeting intimate and dramatic scenes, the group splits up and is decimated. Adawale (Eriq Ebouaney), one of the two Nigerians, is attacked by his brother Bola (Doudou Masta) – who was infected and transformed into a zombie shortly after abandoning the group – and is forced to kill him. The police officer Quessem (Jean-Pierre Martins) faces the zombie horde to save Aurore (Claude Perron) and the baby she is expecting.

They alone will leave the building unscathed, when the sun is already high. But the final message is clear: even if they likely are the only human beings left on earth and of opposite sexes (which implies the possibility of a new postcolonial and mixed society), Aurore (the name now absolutely evocative) kills Adawale to avenge her partner. She is the ultimate girl – a figure so important in horror movies, from Laurie Strode in John Carpenter's *Halloween* (1978) to Sally in Tope Hooper's *The Texas Chain Saw Massacre* (1974) to Sydney Prescott in Wes Craven's *Scream* (1996) – who dramatically expresses her hatred and suffering for the loss of her partner and, although pregnant, never withdraws from the fight. Even in the face of the extreme threat of disaster and end of humanity, the fantasy of white supremacy does not make allowances. Despite the fact that the forgotten colonial past has re-emerged, requiring alliances between new victims and executioners of post-colonial France, the future is a white she-warrior and her baby.

In Danny Boyle's *28 Days Later* (2002) two people of opposite sexes and different colours survive, envisioning the creation of an unorthodox mixed-race family that offers hope for a new world 'purified' of racialised and gendered violence. The survivors are an Irish man called Jim (Cillian Murphy), a young, black woman called Selena (Naomi Harris), and a white working-class child, Hanna (Megan Burns), whose father Frank (Brendan Gleeson) was infected and shot by the military before his mutation. In this film, the apocalyptic scenario is triggered by animal rights activists liberating monkeys infected with a virus that triggers murderous and cannibalistic instincts. The activists and the laboratory staff are all infected, initiating the epidemic.

The film starts with a simple incident: the protagonist, Jim, is hit by a car and falls into a coma. He wakes up 28 days later, in a London that is seemingly abandoned, only to discover that, in fact, the city is far from dead but inhabited almost exclusively by infected people. Jim will be rescued by two survivors, including Selena, with whom he will travel through the rubble of a world that still offers hope for humanity and salvation. They eventually reach a fortified mansion, held by the military, but there they meet further violence as the military want to rape the two girls and show no mercy towards the zombies (represented here as sub-humans, transformed by the infection into rabid animals). In fact, Major West, who is in command of the battalion, keeps an infected soldier, Mailer, in the courtyard of the building in order to find a 'cure':

> West's decision to let Mailer live is revealing in terms of the biopolitical sovereign's authority over life and death. West decides to keep Mailer alive because observing him informs West about the nature of the 'living dead'. Mailer thus becomes an embodiment of the cultural other – an object for the colonizer's gaze, at once fascinating and disgusting. It is interesting to note that Mailer is portrayed by a black British subject which makes the meaning of the 'other' here, doubly loaded. The discourse of race also operates in the manner in which Mailer is restrained – his hands and feet are not cuffed, but he is chained by the neck; this manner of restraint recalls images of black slavery in which captured slaves were chained by the neck during labour. Ironically, it is through Mailer that the destruction of the rest of the contingent takes place later in the film.
>
> (Baishya 2011, p. 11)

Released in 2002, seven years before *The Horde*, *28 Days Later* introduces a new type of zombie who is no longer a returning dead, but rather the example of a new and different kind of animalised humanity, governed only by the rules of the strongest in the pack. This zombie is the result of an epidemic (rampant but potentially circumscribed or circumscribable or curable) and embodies the image of both mutants and rabid animals. Unlike Romero's living dead, they do not stumble but run fast, implying a type of film direction that expertly blends action and war movies. A typical European product, these fast-running zombies evoke the idea that only young people can survive (and men particularly), especially suburban ones and, in any case, people who are familiar with weaponry (such as the military). Europe's hope for the future apparently builds on the abandoned people of its past of racial and gender discrimination. If we look closer, we notice instead that these 'margins', in order to be successful in their struggle for survival, need to be reorganised in a traditional gender and hierarchy-based nuclear family that sees the black Selena revert to women's traditional roles of care-givers (she takes care of Hanna) and subaltern companions to their men, and a white man (Jim) who used to be defenceless and powerless (naked, wearing only a surgical green coat) become the new warrior and defender of the two women (on the socially conservative fantasy behind the imagination of disaster, see Broderick (1993) in response to Sontag (1965)).

There are some important contextual narrative differences between *28 Days Later* and *The Horde*. British insularity, at the centre of its national and imperial conception, allows us to imagine that outside, beyond the seas and oceans, the infection has not spread. In contrast, the Paris that we see engulfed in flames in *The Horde* refers to the no longer radial and potentially limitless relationship between postcolonial centres and peripheries, where there are no barriers to the spread of cannibalistic violence.

In its opening scene, *28 Days Later* contains a criticism of humanity's unbridled sense of omnipotence over nature (reason, technique and language vs. feelings and nature). Monkeys, a usual test animal for biological, genetic and chemical experiments, are also an exotic species, the closest to humans, and connected to Western history by many colonial and racist historical events and dynamics. They are not rabid, but they become such after witnessing the 'ultraviolence' (to quote Stanley Kubrick's 1971 masterpiece, *A Clockwork Orange*) of humans against humans. Apparently, they are repeating, reproducing, revenging or simply returning that violence to humans. In *The Horde*, the origin of the resurrection of the dead is not even hinted at, implying that the catastrophe could be one with apparently no meaning, origin or *telos* (aim). Human beings are faced with the end of *humanity as the 'we' knows it* simply because they have been – and still are – as bad to each other *as they were in colonial times*.

In these two films, the catastrophe is set in the urban space of the working-class protagonists (the centre of London in *28 Days Later*) or a space in which they are partly outcasts (the Paris of *The Horde*). The European metropole – the past centre of colonial power, a place that attacks immigrants and has become the testing ground for governmental practices in the present – is the stage for the action. In both movies, the urban mobilisation/dislocation of bodies under the rules of biopolitical control and the collapse of the market occur on the arrival of the cannibals, whose frantic group does not admit barriers of class, race and gender. It is a destructive and engulfing tide that cancels the spatial segregation and devices of capitalist production embodied in European cities and border regimes separating Britain from the rest (in one of the first scenes Jim is told by Selena that the day before all radio and television broadcasts stopped, there were reports of infection in New York and Paris) and separating the rest within Britain from its privileged top rank (a desert Westminster bridge is the one Jim crosses after leaving the hospital, where he sees sterling banknotes lying on the ground).

Fresnadillo's *28 Weeks Later* is the sequel to *28 Days Later*: after the infected begin to die of starvation, NATO forces take control of Britain. Twenty-eight weeks after the outbreak, an American-commanded force brings in the settlers. A quarantine area that allows the Army to apply the 'let them die' biopolitics reminds us of modern and postmodern camps, left with no aid and hopeless. But it also has strong connections to the postmodern chronotope of the boat that is often left in distress in the Mediterranean – thus increasing the risk that migrants die of starvation, burns, thirst, epidemics, heat and cold as the result of a 'necropolitics at sea' (Casid 2018) that has transformed the Strait of Gibraltar into the 'world's largest mass grave' (Ajana 2013, p. 58; see also the video *Liquid Traces: The Left-to-Die Boat Case* by the Forensic Architecture collective at Goldsmiths College, 2014[8]). In fact, as Baishya (2011) and Bishop (2010) remind us, unlike *28 Days Later* – which recalls the horrors perpetrated both in the colonies and within totalitarian regimes (Pol Pot in particular) and the fear of epidemics – *28 Weeks Later* directly references the war on terror. The only safe area for humans after the quarantine is the 'green zone' in central London called District 1, like the International Zone of Baghdad that was taken over by U.S. troops in 2003 during the Iraq War. The epidemic returns, and with it the *fast-running* and *devouring* infected post-humans, when Alice (Rose Byrne), the wife of the protagonist Don (Robert Carlyle), is found alive in their abandoned house in the quarantined zone by their two children, Andy and Tammy. She was supposed to be dead, left to cannibals by her husband, who managed to reach the Army headquarters. She is a healthy carrier of the virus (as is her son). When the Army abducts her, she is hosed, scrubbed and restrained against her will, thus crossing the quarantine borders and exiting the 'state of exception', the camp, that has turned the infected into *zoe* (as opposed to *bios*, that is, social life) left to starvation, 'bare life' in the hands of the military. Because of the fragility of the camp's boundaries, the quarantine here recalls past colonial histories:

> Enclosure, gated enclaves, and sequestered populations produce imagined and real threats—transgressions of the private, intrusions on the safe, a storming of the gates. They take the form of enclosure and containment, but what and who must be kept out and what and who must stay in are neither fixed nor easy to assess. Internal enemies are potential and everywhere. To protect those within or contain them, or to protect those outside who might be disturbed, *at risk*, or endangered by exposure, were not mutually exclusive projects. Nor were, and are they, opposing ones. Being 'at risk' and 'a risk' is a fuzzier political line than most colonial histories allow us to imagine. Being protected is not designated only by who is within the colony and who is not. This quality of and quest for boundedness within and outside produces both its desired and unwelcome effects: transgressions, escapes, flight, detention, suspicion, illicit border crossings, entrapment, and more surveillance. But who is the 'hunter' and who is 'hunted' are not fixed ontologies in this furtive, bellicose space, but unstable ones. Who is 'hunted' may turn into the hunter for despite the dissymmetry of material force and resources, the installment of fear can work the other way around.
>
> (Stoler 2016, p. 121)

This unfixed ontology is reflected in the blurred distinction between the human and the rabid post-human: in *28 Days Later*, when confronted with violence and rape at the hands of the military at the Army 'camp' near Manchester, Jim:

> can't avoid or control his innate, personal rage. The effect in either case is practically identical: for all intents and purposes, in the moments of his rage, Jim is [a] zombie [...] He sides with the zombies to kill his human tormentors, and even the zombies can't tell he's not one of them.
>
> (Webb and Byrnand 2008, p. 86)

In *28 Weeks Later*, once Don reaches Alice and kisses her, he immediately turns rabid, infecting all the 15,000 civilians who had survived the epidemic. But unlike the other rabid post-humans, he is after his son: zombies are not 'devoid of history' nor memory. They seek revenge – apparently – or a way to survive, killing the only humans who could help other humans defeat the virus. Paradoxically, the sacrifice of many soldiers to rescue the child and take him and his sister where he could be studied and a cure found leads to the spread of the virus outside Britain, as the last scene in a zombie-ridden Paris reveals. In this film, similarly to *WWZ*, scenes of the massacre of the rabid post-humans recall or accompany the extermination of the survivors, as if to imply that the 'state of exception' may affect everyone who is not protected by (military) power. While the 28-day quarantine imposed in London recalls the navy's or coast guard's failure to assist migrants in distress at sea, resulting in their death in shipwrecks, the clean shot that exterminates all human survivors to stop the mass infection caused by Don in *28 Weeks Later* may be seen as an allegory for the indiscriminate bombing of Iraq and Afghanistan in order to stop terror attacks.

In *28 Days Later*, *28 Weeks Later* and *The Horde*, power is described as devoid of *love, empathy, respect* and *loyalty* to people and relationships. Soldiers, with very few exceptions, can easily turn into emotionless killing machines (e.g. the slaughtering of civilians in District 1, in *28 Weeks Later*), recalling the 'zombification' of human soldiers in Kubrick's *Full Metal Jacket* (Nì Fhlainn 2011, p. 143). Humanity is still depicted as bound together by feelings, but it is doomed to extinction at the hand of the rabid monsters or the military – that is, at the hands of the horrors from the past resurfacing in the present or at the hands of the institutions that are perpetrating such horrors in the present and are sometimes indistinguishable from the rabid post-humans. Here humanity comes to terms with itself, with its own vanity and fear, selfishness and hatred. The easiness with which the military and civilians alike can turn into zombies, as shown in *28 Days Later*, *28 Weeks Later* and *The Horde*, makes us question Elias Canetti's argument that the representation of death 'liberates' the living spectator:

> Fortunate and favoured, the survivor stands in the midst of the fallen. For him there is one tremendous fact: while countless others have died,

> many of them his comrades, he is still alive. The dead lie helpless; he stands up right amongst them, and it is as though the battle had been fought in order for him to survive it. Death has been deflected from him to those others. Not that he has avoided danger; he, with his friends, stood in the path of death. They fell; he stands exulting.
>
> (Canetti 1981, p. 228)

There is no liberation. On the contrary, the 'revelation' the apocalypse brings with it is that everybody is under attack and humans, when they survive, can be worse than their zombified peers. In such circumstances, the very idea of a future seems impossible; a project for the future, after the apocalypse, seems unlikely.

6 Of progress and its reversals

A critique of the superpower of human technologies and their impact on the planet is the underlying theme in *WWZ* as well as in *28 Days Later* and *28 Weeks Later*. In *WWZ*, the epidemic is not due to an apocalypse as a transcendental event but to the work of nature in response to humanity's nefarious exploitation. It is no coincidence that the opening credits are accompanied by TV and radio voice-overs reporting on the impact of man on the planet in terms of global warming, epidemics and pandemics. As in *28 Days Later*, *28 Weeks Later* and *The Horde*, which address the metamorphosis of the undead into infected mutants, the undead here do not rise from the grave but became so as the result of an infection that spreads through a bite, or through biological contamination from a misdirected experiment with the measles virus. In this post-apocalyptic zombie film where the epidemic starts in Pakistan and spreads to Western metropoles, the former UN investigator Gerry Lane (Brad Pitt), is called back into service by international authorities to deal with an emergency comparable to those he experienced in Liberia, Sudan and Palestine. He left the United Nations because he wanted to be with his family – his wife and two little girls – but now he needs to break his promise and again be a 'hero of world peace' and survival. His world tour in search of a solution to the epidemic (South Korea, Palestine, England and the United States) is desperate, and he loses track of his family, first rescued by a U.S. navy carrier in the middle of the Atlantic under the protection of Lane's former colleague and friend UN Deputy Secretary-General Thierry Umuntoni, but then abandoned in an isolated refugee camp because of overcrowding.

Lane flies to Israel, after hearing from a CIA agent in South Korea that the Mossad has better information about the epidemic. While observing the cannibal masses, Lane notices that zombies do not attack those who are sick and suspects that this is to avoid eating infected and corrupt beings. The narrative thus invites the audience to share in an ideology of progress: the survivors (stronger, fitter, and Western), once rescued from the disaster and made even bio-physically stronger through natural selection, will be able to definitely overcome nature's ultimate revolt against human despotism. As he and an Israeli female soldier, Selden, flee Jerusalem (under attack by the infected coming from the 'other side of the wall', so presumably former Palestinians), Lane tries to kill a group of zombies who were hiding on his plane. The plane crashes in the middle of Cardiff, near the World Health Organisation headquarters. Here, he finds test tubes containing viruses and bacteria and is able to confirm his theory and find a cure for the zombie epidemic. In most of its recent output, Hollywood reframes the ultimate human catastrophe as something to which humans are able to find a solution. Mutation seems to be an allegory for the defeat of humanity in its 'war of domination' over nature, but defeat is only temporary: *WWZ*'s apocalyptic scenario turns into a progressive epic where the solution to what was meant to be the end of the world is found by injecting all military personnel with a deadly virus in order to exterminate the rabid monsters.

In *WWZ*, there is no mention of the many physical walls and social boundaries, unbalanced power relations and racist segregations that divide the world's population: the white man's Western-centred fantasy of being able to control nature and catastrophes is finally projected onto the entire world. In this sense, *WWZ* thematises one of the core topics that I will develop in Section 3 of this book in the frame of a critique of the Anthropocene: Western culture's (wherever it applies, even outside the West) anthropocentric and technocratic blind faith that science and technology can solve political and social problems that politics not only is unable to solve but has itself created. This theme contrasts with the portrayal of catastrophe in the other films in this section, in which there is no lone hero, no solution to the crisis, no human revanchism and the characters are eventually overwhelmed by the post-human version of their repressed past and ongoing responsibilities.

When compared to *WWZ*, the three films I discussed earlier in this section share another commonality. This concerns what Talal Asad (2007), talking about the 'little (neo)colonial wars' of the twenty-first century, refers to as the differential space assigned to someone's mourning for those who have become undead, which is dependent on the position of the object of mourning within gendered, racialised and classed global hierarchies. The anonymity assigned to the mob is common to the genre and to the films I have discussed here, but it is less evident or more gently managed in *The Horde* (especially in the relationship between the living Adawale and the undead Bola), in *28 Days Later* (in particular when the father of the little girl notices that he has been infected and gives her his last farewell), and in *28 Weeks Later* (in the military doctors dealing with the infected Alice). It is, however, absent in *WWZ*, where very few scenes address the emotional consequences of the transformation of 'loved' ones or acquaintances into cannibal persecutors. While fleeing a 'lost' Philadelphia under siege, the Lanes are hosted by a Latino family who are later infected. Their child

Tomas manages to escapes and join the Lane family on the roof of a building but is never seen mourning the transformation of his loved ones – not even in the scene where his father and mother are trying to tear him to pieces. Instead, the emotional and intimate tension deriving from the fear of loss (from mutation and thus death) is exclusively reserved for Gerry Lane's very white family. Indeed, his concern for his wife and daughters, who are left alone to deal with the war economy and the U.S. Army's 'refugee management' policies, is a main theme in the movie. Unlike other innovative plots, in which courage is not the exclusive mark of white heterosexual men and fathers, female roles in *WWZ* are ancillary to Lane's actions – as is the case with Selden, the Israeli female soldier he saves by cutting off her infected hand.

The emotional element is thereby so completely reduced to the protagonist (and to his family only by extension) that some of the scenes portraying survivors' joy risk appearing grotesque and significantly out of place – such as the scene in which many Jews and very few Palestinians sing, while outside Jerusalem their loved ones devour their relatives, former colleagues, neighbours and friends. The existence of a common enemy allows a coalition to form between those who had hitherto hated each other. But those who have remained beyond the wall, in that area normally known as Palestine, are the Palestinians who have been confined for decades in liminal areas afflicted by poverty and lacking any means of defence – except for self-organised armies. In this sense, I agree with Luckhurst that the film's representation of zombified Palestinians attacking Jerusalem is one of racially demonised monsters (2015, p. 186). Here, the walls built to separate the occupied territories from Palestine have been reinforced, multiplied and made even higher to neutralise the zombies. But the barriers built by men can do nothing against the fury of the monsters/terrorists, whose threat recalls the 'natural' condition of permanent fear of extinction of Jerusalem's (Israeli) inhabitants. The monsters/terrorists are like a sea tide in the film, where Marc Forster, who must have been influenced by Bayona's *The Impossible* (2012) and its admirable depiction of the devastating force of the 2004 tsunami, shows them running to build a pyramid of flesh that scales the protective wall onto fleeing humans. The aerial images of the infected, bloodthirsty, and anonymous tide greatly resemble the shot from above of the violence and destructive speed of water. One might even compare these effects to aerial shots of fleeing civilian populations in war films that made history and created a model script for those that followed, such as Ralph Nelson's *Soldier Blue* (1970) on the Wounded Knee massacre as a memory of the future My Lai massacre in Vietnam (1968), Francis F. Coppola's *Apocalypse Now* (1979) and Stanley Kubrick's *Full Metal Jacket* (1987) on the Vietnam war.

The allegory of the camp here – which is not 100 per cent fictional as in the case of *28 Days Later* and *28 Weeks Later* but refers to the existing wall dividing Israel and Palestine – is used to claim the camp/colony as the only space where humanness, humanity and civilisation can be preserved from barbarity, decay and death (Stoler 2016, pp. 37–67). Camps and colonies, together with carceral archipelagos and borders, are not only biopolitical *dispositifs* but also the symbolic referents that have been structuring and regulating Modernity since colonial times (Stoler 2016). Today, the fictionality of *Fortress Europe* recalls the idea of postcolonial Europe as a camp – as much against migrants and refugees as against epidemics and all side effects of what is meant to stay outside (wars as well as environmental disasters). Here, it is not only the idea of everything outside Europe and the West as being corrupting that is colonial in itself, but also the belief that lethal epidemics are the result of contact with the uncivilised and the poor – or even the outcome of the revenge of former enslaved or colonised peoples who, as they head 'northbound' (as if they were only migrating to the North), bring death to the former colonial metropole. As Roger Luckhurst reminds us:

> The tie of viral epidemics to the history of the zombie was consolidated in the early years of the HIV crisis, when a high preponderance was noted in the immigrant Haitian population in New York and it was called 'an epidemic Haitian virus'. Under the headline 'Night of the Living Dead', the *Journal of the American Medical Association* speculated, entirely fantastically, that HIV was spread by Voodoo rituals using human blood. This simply rehearsed the age-old tactic of marking the migrant and foreigner as fatal disease invader, the demonized carrier, but the modernity of 'emergent infections' in particular resides in how they always trace out colonial and post-colonial pathways.
>
> (2015, p. 181)

The colonial trope, which I will discuss in greater detail in the third section of this book, is deeply embedded in *WWZ*. The world is represented as divided into places where (bio)technology governs nature (that is Europe and the West, including Israel, although these places are not internally uniform) and places where the barbarians have no technological capacity. According to this trope, disasters perceived as endemic to barbarity become a catastrophe only when they spill into the space of rationality and technology: it is not by coincidence that a state of emergency is called when the epidemic reaches the United States.

The allegory of the camp/colony is very much at the core of *The Walking Dead* (2010–) The saga, which has now reached its tenth season (an eleventh season was recently announced) has always revolved around a few main themes: the need for survivors to band together for self-defence, the search for a safe space to escape the zombie mob and sustain life, the need to fight wars for supremacy against other survivors and the need for a strong leader – a sort of human Leviathan (Lucci 2014, pp. 112–113) – who can coordinate defence, movements, supplies and war. All these themes are deeply embedded in the colonial history of the United States: zombies (the result of an infection that turns the mortal living into immortal dead) can be seen as an allegory of revenant post-human indigenous populations who are everywhere (as it used to be) and in a sort of historical reversal force humans into makeshift reservations. As we have seen, according to the seventeenth- and eighteenth-century British colonial discourse that was integrated into the United States Constitution and political culture, the indigenous peoples, just like zombies, were considered unable to master nature and make the land productive. Zombies may then represent the new silent colonisers, like the migrants from Latin America described by the news media and other institutions as a dangerous unruly mob that will reproduce in great numbers, changing the racial composition of the country: against them, walls and fences need to be built and colonies resurrected. Or they may be an allegory for the victims of the violence of slavery, given the location where the TV series is filmed, Georgia, with its history of plantations and the role it played in the Civil War. But the real monsters here are the humans, capable of slaughtering each other in the name of supremacy, honour and revenge. All this violence – which brings to mind the Hobbesian 'state of nature' (Lucci 2014, p. 113) – is not aimed at building a future without epidemics or a planet inhabited by healthy humans. In fact, all humans are already infected: the threat that was seen as coming from outside is actually inside 'us'. When humans die, they turn into zombies. Thus the existence of *humans* is very limited in time. The *zombie apocalypse* is a *total* human catastrophe which references the war on terror in both the limitless effects of weapons of mass destruction and the potential infiniteness of the war, constantly echoed after 9/11 in discourses legitimising Western military intervention in the Middle East. In fact, the repetition and circularity of contagious diseases, death, wars, and creation/destruction of shelters/camps – as if there was no distinction between the past, the present and the future – can potentially take place *ad infinitum*. By staging a potentially never-ending repetition of the same events (which makes colonisation by humans *impossible*) and the constant return of colonial violence (land occupation and massacres), *The Walking Dead* reverses the Enlightenment's idea of progress. As argued by Reinhart Koselleck (2004), the idea of a linear and positive progression towards perfection or, in the words of Jeremy Bentham (1768), towards the 'Greatest Happiness of the Greatest Number' was at the core of Modernity. Humanity has jumped into the past and is stuck with it, as the colony will not last and the State cannot be built: all borders are temporary and weak, security at the frontier is ephemeral, and internal governments cannot resist in the face of human disruptions, conflicts, fear and death, mainly because there is no clear distinction between the infected humans who must be defended and the walking dead who must be killed. The only difference lies in the social organisation of life (*bios*) and post-human *zoe*, a separation that, again, is absolutely temporary.

In Francis Lawrence's *I am Legend* (2007), the post-human mob is more one of mutants than of zombies. Loosely based on the 1954 novel by

Richard Matheson, the film follows its first two adaptations, *The Last Man on Earth* (1964) by Ubaldo Ragona and Sidney Salkow, starring Vincent Price, in which post-humans are more vampire-like, and *The Omega Man* (1971) by Boris Sagal, starring Charlton Heston, in which the threat comes in the form of albino mutants. In the film, unlike the novel, the only survivor is a U.S. Army virologist, Lt Colonel Robert Neville (Will Smith), who lives segregated with his dog in the apparently deserted city of New York after a modified virus created to defeat cancer has turned lethal, killing 90 per cent of the world population and turning most survivors into vampiric mutants who hide during the day to avoid sunlight. In the novel, Neville has no science background and decides to get the knowledge he needs to find the cause of the epidemic – a bacteria. In Lawrence's film, humans turn into mutants because of a fever for which the virologist, after capturing one of them and keeping her in his own laboratory, finds a cure. There is no cure in the novel, however, where a new mutant 'race' – neither vampiric nor human – will take over, exterminating both the older race, including Neville and the monsters, and building a new society. In the 2007 film, Neville self-immolates to let the other two survivors, a woman and her child, reach a survivors' camp in Vermont. As in *WWZ*, the only hope here is the ability of the First World (and the United States in particular) to master biotechnology and military knowledge, an assumption that is completely absent in the pessimistic narrative of *The Walking Dead*. Like in *WWZ*, there is no engagement with the zombies: it is the context (the quasi-extinction of humans) in which Neville lives that pushes him to find a way to reverse the mutation. The context in which Lane operates in *WWZ*, is instead one where a functional army is still in place and many humans are hiding (including his own family), and annihilating the enemy is still a possibility. Finally, unlike *WWZ*, where the catastrophe is due to the enemy's 'bad will', in *I am Legend* it is assumed to have been caused by human arrogance. In all these films, the enemy – who used to be one of 'us' and is still similar to humans, albeit reduced to his/her *devouring* animality (Coulombe 2012) – is an allegory of the invasion from within and without, from a distant geography and a history of violence. In *The Walking Dead*, proximity to the zombie (as one of 'us') is more deeply thematised, although the condition of zombie is not reversable. In fact, the issue of resurrection evokes the biblical apocalypse that in Christianity corresponds to the return of the 'wretched of the Earth' for reparation or revenge. In *The Walking Dead*, zombified friends and family trouble the daily life of the survivors, as if they were coming back, physically or in their dreams, to remind them of their fate. In the midst of a de-humanised, de-individualised and massified mob of dispensable beings, they seem to remind the living that everyone is already infected and sooner or later will be like them: the difference between humans and post-humans is thus ephemeral, as humanness is only a temporary state of post-humanness.

> Our fascination with the monstrous has a great deal to do with our very uneasy sense that such creatures are forms of ourselves – the human gone terribly wrong. The monstrous thus points to our understanding of the precariousness of human identity, the idea that human identity may be lost or invaded and that we may be, or may become, something other than we are: monsters have something to show us about our world and ourselves.
>
> (Neocleous 2005, p. 5)

This is why one of the characters refuses to kill zombies: not only do they remind him of his nearest and dearest but also trigger the sense of guilt he feels for something that has occurred in the past. The violence the survivors have been inflicting on them is a foundational violence that humans will pay for. Next, I will connect representations of rabid post-humans, terrorists and migrants to analyse proximity and distance, mourning and indifference, compassion/empathy and violence, which are deeply interrelated in the Western imaginary of the monster.

7 Of rabid post-humans, terrorists and migrants

In all the films I discussed above, zombies figure mostly as an anonymous mob with no social connections, emotions, aspirations and consciousness, and reduced to animality (their only functioning organ being the hypothalamus, which regulates hunger). They constitute, as Maxime Coulombe notes, 'a *negative* specie, as if they were the cursed side of the human being, made all of a sudden *visible* and *vengeful*' (Coulombe 2012, p. 65, my translation). Zombies can and ought to be sacrificed in the name of humanity and civilisation, for they are 'living' what is considered an unworthy 'life', akin to a post-human version of the Greek *homo sacer* in Giorgio Agamben's interpretation (1995). But there is another important feature connected to their expendability: the different time dimension they live in, which is one of infinite repetition and suspension. In brief, a feature is inscribed in their nature that in the human world is inflicted, in a variety of ways, on subjects of unruly mobility:

> camp and prison sites proliferate. Transfer and deferral are the strategies of the broader closure problem that is the carceral state. That closure cannot come to a close but must take the shape of transfer, extension and repetition is not exceptional; it is integral to the state of exception that is the ongoing refugee crisis that enwraps the terrorist and the migrant, the refused refugee, as terror-figures. To transfer is to extend the temporal illogics of indefinite detention in the refugee camp and the prison camp across an undefined and widespread global geography.
>
> (Casid 2018, p. 211)

This is what, in the case of the hotspot system, Glenda Garelli and Martina Tazzioli (2018) call a system of 'flexible checkpoints of mobility disruption' for the containment and channelling of migrant mobility beyond temporary detention, a system that creates potentially infinite opportunities for bouncing migrants around.

As Shahram Khosravi argues for the deportation regime, which is a side operation of the border regime,

> in the condition of 'illegality', [...] time has a different meaning and rhythm (see Willen 2007). Ehsan, a young undocumented Kurdish man in Stockholm, hated all weekdays and most of all Mondays. In his view, on the first day of the week all 'ordinary' people started a meaningful week of work while he did not: Mondays represented 'moving forward', while he just 'remained at the same point'. His time was not that of 'ordinary' people. Many other informants also used terms like 'dead time' or 'a time of death' when talking about their lives; they did not consider their time of hiding as part of life but rather a time of non-existence.
>
> (Khosravi 2010, p. 91)

> One feature of becoming 'estranged citizen' after deportation is a sense of 'not being in-time with others' or a sense of non-simultaneity that emerges from the lack of 'the sense that others are doing at the same time things that are meaningfully related to your own experience' (Boyarin 1994, p. 17).
>
> (Khosravi 2018, p. 7)

Zombies dwell in a condition of infinite repetition that reflects the experience of both the illegal migrants caught in the hotspot system and the deportees and recalls what Johannes Fabian (1983) argued about the different temporality of the colonial Other, that of being stuck in a past, the 'waiting room of history' in the words of Dipesh Chakrabarty (2000), and doomed to an infinite temporal loop. They bring (the temporality of) death to the Earth, and as such force the life of the living into the repetition of self-defence mechanisms. Fears of death, physical and emotional, are connected here to fears of social death – a state where there is no hope for the future because, apparently, there is no future.

These essential features of the zombie, regardless of direct or indirect references to the war on terror, are what makes the post-human mutant an allegory for the mainstreamed figure of de-humanised global terrorist fabricated in the West that Talal Asad (2007) brilliantly deconstructed in his book, *On Suicide Bombings* (see also Fekete 2009; Furedi 2007; Deltombe and Rigouste 2005, pp. 195–202). This figure is often depicted in military dispatches, mainstream political debates, and international geopolitical agreements as moved by an 'uncontrollable anger' (rabid, indeed), a lack of hope in the future, and death impulses. It is also described as lacking empathy towards human life – rather than embodying an 'act of death-dealing that reacts to injustice by transgressing the law' (Asad 2007, p. 47) – and, for that irrationality, deserving only to be killed:

> The right to kill is the right to behave in violent ways toward other people – especially towards citizens of foreign states at war and toward the uncivilized, whose very existence is a threat to civilized order.
>
> (Asad 2007, p. 60)

This opposition between the civilised order on the one hand, and (potential) terrorists whose life is considered worthless on the other, reflects the idea of two separate worlds informing counter-terrorism handbooks that Louise Amoore and Marieke De Goede have described as:

> Two worlds of globalisation are represented through risk practice in the war on terror: one populated by legitimate and civilised groups whose normalised patterns of financial, leisure or business behaviours are to be secured; and another populated by illegitimate and uncivilised persons whose suspicious patterns of behaviour are to be targeted and apprehended.
>
> (Amoore and De Goede 2008, pp. 5–6)

The post-9/11 geopolitical landscape is described by Asad (2007) and Mbembe (2003) as driven by neocolonial attitudes. Their vision problematises Foucault's opposition between two different modern biopolitics, namely the disciplining strategies of 'killing/letting live' (sovereignty) and 'fostering/disallowing life' (disciplining biopolitics) (1976). Asad and Mbembe argue that neocolonial situations see the coexistence of both aspects and are based on the fundamentally colonial attitude of classifying Others as 'expendable'. The neocolonial lack of responsibility for the victims is structural to a 'war of the worlds' setting where the enemy is not addressed as a nation (that is, as an entity deserving of all rights determined by international law), nor as a human being (in the case of zombie narratives), but as a *devouring* monster.

The 'killable' Other is the product of a conception of 'life' within the 'frame of war' (Butler 2009), where 'small colonial wars' (Asad 2007) are engaged against the monster in order to spread freedom, progress and peace around the world. Therefore, insofar as military interventions by Western powers continue this colonial tradition, it should be evident that their primary aim is not the protection of life as such but the construction and advancement of specific kinds of human subjects and the outlawing of all others.

In the frame of those 'small colonial wars', transnational migrants are among those paying the price for an interpretation of society that externalises – as with the Paris attacks – its own foreclosed colonial memories and considers such acts the result of external terrorism. The supposed state of permanent crisis determines, and is said to be determined by, a constant influx of migrants and the alleged 'persistent risk' of *risky bodies*, meaning old and new migrants and their offspring, turning into members of terrorist cells. This potentiality allows and justifies re-allocating power to the State, reinforcing its power to set a hierarchy among people, nationally and internationally, along colour, gender, religious and class lines. The definition of *risky body* comprises two meanings: the *risky body* and the *body at risk* (Aradau 2004). *Bodies at risk* and *risky bodies* are on the same boat, because they were conceived either as victims of a disaster or as threats and/or vectors of disaster into the European space. This iconography, which reveals postcolonial global inequalities, is reactivated when assigning a victim role to the women and children who cross the Mediterranean while the young and adult men are criminalised. (This specific element of the discourse on migrants and refugees is currently investigated by the research project '(De)Othering: Deconstructing Risk and Otherness: Hegemonic Scripts and Counter-narratives on Migrants/Refugees and "Internal Others" in Portuguese and European Mediascapes', 2018–2121) The double (gendered) feature of the *risky body* is attested in securitisation studies (Buzan et al. 1998; Bigo 2002; Huysmans 2006), which have shown how migration can be turned into a security issue that the humanitarian rhetoric can help sustain as an essential element of governmentality, especially with respect to governing human mobility (Bigo 2002; Agier 2008; Ticktin 2016; Tazzioli and Walters 2019). My analysis builds on these results to move towards an understanding of how *bodies at risk* and *risky bodies* are conceived as jointly infused with a *dormant barbarism* that makes the victim a *potential monster*. A state of victimhood can thus always be reversed back to one of inherent danger, while security measures, manuals and practices ontologise, de-politicise and de-historicise the condition of migrant/refugee. The ontologised condition of the *risky body* is constituted by both inferiority and its 'illegitimate' mobility. By ontology of inferiority, I mean the outcome of a discourse that:

> concealing the migrant's history, by abstracting the migrant from his or her immediate social context, the migrant can then be repositioned outside modernity and into what [Anne] McClintock (1995) calls 'anachronistic space': a space imagined as 'inherently out of place in the historical time of modernity.' In colonial thought this has normally meant casting the Other as prehistoric, savage, and atavistic [...] the Other is displaced onto a rather unique posterior, future-conditional time. [...] Not an ontologically discrete future time-space that comes after the present in an ordered chronology, but one situated between present and future in which the future is already present in the present, albeit in virtual form.
>
> (Baldwin 2013, p. 1483)

By illegitimate mobility, I mean instead the idea that racialised Others escaping the control of both local state institutions and international organisations that want them sedentary (Bakewell 2008; Castles 2010; Bettini 2013) make themselves into a problem. Further, when attempting to enter Europe's space, they are seen as violating the ban imposed on (post-)-colonial subalterns to prevent them from crossing the border that constitutes them as such. In fact, as a result of the border's semiotic power as a biopolitical *dispositif* – which I will discuss in more detail in Section 2 – the illegalised and threatening 'body in motion' becomes subaltern. The symbolic and material camp of power relations in which this designation takes place is sustained by colonial and cultural (national) archives of figures of race, which rearticulate old racist meanings and produce new ones. It is within this semiotic landscape that the ontologised construction of migrants/refugees and alleged terrorists as monsters takes place (or, in zombie narratives, the external threat of the *fast-running* and *devouring* post-human and the will-be-rabid monster among us). These figures simultaneously threaten society and confirm 'which' society should be defended (Foucault 1976). In the frame of the war on terror, the charge of terror reinforces this ontology, adding the ferocity of blind fanaticism to the picture.

When defining who can be considered 'torturable' and 'expendable', 'excluded', 'included', or 'partially included' within an 'encamped space' (Gilroy 2000, p. 83) defined by social and political rights and reproduced by its own 'deportation regime' (De Genova and Peutz 2010), the line is drawn at the ontologised figure of *risky body* – the migrant, the terrorist, and the rabid post-human.

Nevertheless, in the zombie narratives I have explored, perhaps because of widespread criticism of the un-human conditions to which alleged terrorists were subjected in 'carceral archipelagos' (Agamben 2003; Stoler 2016) such as Guantanamo and Abu Ghraib (Baudrillard 2005; McClintock 2009; Butler 2008), or in the migrant prisons of Libya and Turkey, an attempt is made to rethink the relation between border, security and war in the light of humanitarianism. Unlike *WWZ*, the British and French productions make room for the idea that, although it does not seem possible to save the monsters from themselves, an ethical approach is needed in order to *stay human*. In order to stay human, *we cannot be like them.*

I believe that there is a link between the humanitarian rhetoric, which sustains a critique of the border regime and un-human practices against alleged terrorists, and the positionality of the spectator of zombie films or images of migrants penetrating Europe. In both cases, the gaze on the *risky subject* – be they a migrant, an alleged terrorist or a fictional running and devouring infected – is not neutral. Those who look are indeed subjects who, in the act of looking, are also affected by what they see and interpret it accordingly. When looking, they connect their past visual and cognitive experiences to what they see and, in the process, select some details over others, neglecting many. One's positionality affects individual and collective experiences of seeing. In the case of visual accounts of past experiences as well as in the case of visual texts, the way in which the 'we' interprets what the 'we' sees is necessarily influenced by many factors, personal and emotional ones included. However, 'our' interpretation is also over-determined by social *scripts*, which through iteration, as John Tagg (1988) and Judith Butler (1993) have pointed out, *produce truth*. As argued by Nicholas Mirzoeff (2011), the spectator's 'right to look' encompasses both the right to look and the right to identify themself in what they see by means of self-representation.

In my interpretation, fears of the unruly appearance and mobility of the rabid post-human are an allegory of the *moral panic* over a deadly pandemic that as such would dissolve the very fundaments of life and society. In fact, rather than personal death, what is represented here is the annihilation of both *bios* (or social life) and the 'civilised' world (or the 'worthiest' portion of humanity, the West, taken as representative of humanity as a whole). In Section 2 of this book, I will explore in more detail how mobility of colonial subalterns and minority subjects has been historically considered as carrying an omen of destruction (because of the social order that constructs it as such). Because they embody the possibility of subverting the material and symbolic order of the status quo, their mobility has accordingly functioned as a trigger for *moral panic*.

The spectator is thus not invited to identify with the monster, the *fast-runner* and *devouring* zombie. Rather, they are invited to keep an ontological distance from the subject who is at the same time *at risk* and *risky* – simultaneously victim of violence (the epidemic, the killing of rabid human beings) and a criminal violating the security border that protects human life (the quarantine in *28 Weeks Later*) and, allegorically, a '*fortressed*' West. In films telling stories of contemporary monsters, the hero is the *subject who looks* at the violated boundaries and borders, and their positionality is accordingly revealed as being essentially oppositional to the one embodied by those who come from across the *human* border. Furthermore, representations of the Other who has violated the boundaries of the human becoming post-human, or allegorical representations of the postcolonial migrant who comes from across the sea, serve a function that I would call *identification by contrast* – that is, the possibility for *those who look* to define who they are or would like to be *against* those who are 'not like them'. The monster is created by the horror we are filled with when looking at the spectacle of border-crossing, where the physical and symbolic border that is crossed is the one separating the Global North from that colonial and postcolonial violence that is pretended to be located over the 'abyssal line(s)' (Sousa Santos 2014, pp. 118–135, 2018). In fact, these are the lines delimiting the symbolic and material space of the *hic sunt cannibals*. Again, we cannot be relieved at watching a monster die, because the apocalypse will be *coming again and again and again*.

Moral panic over a monster 'invasion' ought to be understood as integral to the guilt the 'we' feels for the unequal power relations along the 'abyssal line' grounded in the constant reproduction of the inferior status of Others-at-the-frontier – including the *abject* status assigned to post-humans. As Coulombe highlights for zombies, they are 'a *negative* species', 'made all of a sudden *visible* and *vengeful*' (2012, p. 65). Their plans for revenge match the spectator's sense of guilt for colonial injustices, which, however, is accompanied by the desire to be absolved of that violence and by a politics of victimhood that prevents them from identifying with the monster (who may be one of us, or someone we once knew, as in Michael Hanneke's 2005 film *Caché*).[9] Rather, it prescribes an ontological distance from the subjects-in-motion, seen as devoid of subjectivity and voice. In this regard, compassion, guilt and solidarity activate discursive dynamics and silencing *dispositifs* that share a dangerous familiarity with the symbolic materials at the very foundation of the civilising mission: humanitarian efforts to 'rescue the good savage' and the white spectator's empathy for the suffering of the black enslaved in the United States, as argued by Hartman:

> The ambivalent character of empathy – more exactly, the repressive effects of empathy – as Jonathan Boyarin notes, can be located in the 'obliteration of otherness' of the facile intimacy that enables identification with the other only as we 'feel ourselves into those we imagine as ourselves'. And, as a consequence, empathy fails to expand the space of the other but merely places the self in its stead. To not suggest that empathy can be discarded or that [John] Ranking's desire to exist in the place of the other can be dismissed as a narcissistic exercise but rather to highlight the dangers of a too-easy intimacy, the consideration of the self that occurs at the expense of the slave's suffering, and the violence of identification.
>
> (1997, pp. 19–20)

The problematic political edge of feelings like 'compassion' and 'empathy' as the engine of humanitarian discourse against colonial and postcolonial violence *at the border* has been discussed by Eric Fassin:

> The tensions between compassion and repression, the problems posed by the mobilization of empathy rather than the recognition of rights, the prejudices toward the dominated and their consequences regarding the way to treat them have a high degree of generality that make them relevant in various contexts. Configurations may be different, but processes are similar.
>
> (2011, p. x)

In a recent study on representations of migrants and refugees in news media and films, I have discussed pictures and films on migrant landings in

Lampedusa (Giuliani and Vacchiano 2019). Films like Francesco Rosi's multi-awarded *Fuocoammare*, among others, are an important contribution to a reflection on *some* aspects of the Middle Passage. Above all, they are testimony to the relation between the *humanitarian reason* and the symbolic and emotional labour of the border regime that governs migration. The spectator's gaze is caught and moulded, guided towards specific interpretations. Ideas and imaginaries that drive these interpretations are the ones producing and reproduced by laws stating that 'they' (the *risky bodies*) are allowed to enter the safe space of *Fortress Europe* only by virtue of their essentialisation as *bodies at risk*, that is, vulnerable and voiceless governmental objects of Europe's border biopolitics and (white) saviours. The *risky body* is allowed to enter only if and when it has been transformed into a docile, less dangerous subject, thus allowing 'us' to see ourselves as benevolent carers, or better, as benevolent controllers, because the threat is never over. In the immunisation process described by Roberto Esposito (2008), an enduring threat is needed to legitimise both border closures and 'humanitarian exceptions'.

Today, images of landings and boats in distress include the widely circulated photographs of a crowded boat arriving at Lampedusa;[10] of the police, coastguard and medical personnel helping migrants disembark from a rescue vessel;[11] and of rows of migrants squatting on the ground at the island's pier.[12]

These images suggest a reading of the 'Other-at-sea' as distant (physically separated from the receiving land by the sea, from island dwellers by the pier and the police line, and from their rescuers by white rubber gloves and hazmat suits),[13] vulnerable (*at risk*), threatening (*risky*), and voice-less. Nakedness, deprivation, desperation: the depoliticised celebration of mass suffering positions the receiving nation, institutions and civil society as heroes against a depersonalised and dehistoricised violence (perpetrated by gatekeepers and traffickers directly paid by or indirectly profiting from the European border system and agencies). Their representation as a voice-less mass of individuals, with whom the 'we' apparently has no intimate or emotional connection, makes me turn my gaze back to zombies and their more common representations.

In the introduction of this section and in Chapter 4, I briefly analysed the logic behind the BBC mini-series *In the Flesh* (2013–2014) – that is, the zombie is 'one of us'. The zombie uprising is defeated by the living, but not through extermination. The undead do not disappear; instead, they are domesticated.[14] This recovered proximity is sometimes seen as a gift, sometimes as horrific event.

The Human Volunteer Force (HVF) – which was not dissolved after the war – thinks otherwise, that all 'rabbits' (the rabid undead) and partially deceased syndrome sufferers (PDS) should be exterminated. The clash between humans and post-humans works as an allegory for the struggle between real humans and their Others, Christianity and its monsters, the West and the rest, at the core of Western ideas of colonial and postcolonial

Modernity restated and revised in the war on terror. Nevertheless, if we read the series through this metaphor, the result would be politically problematic. What seems problematic is using the cannibalistic undead as an allegory for the (internal/external) Other in British society, as well as attempting to 're-humanise' it through the benevolent Western and civilising 'medical technique'. There are disturbing parallels between such social dynamics and extreme *Othering* of what is non-'normative', which I have described in the case of European Jihadists and illegalised migrants.

> In seeking to extend 'grief and care to the dead stranger', [public mourning, liberal hospitality, and calls for multiculturalism] not only transform the migrant into a predetermined universalised figure in need of Europe's help and hospitality, they also reproduce a narrative of European goodness and benevolence.
>
> (Danewid 2017, p. 1682)

Emotional ties between the inhabitants of Roarton and the undead are only possible after the undead are declared victims (i.e. meaning that their violence when they were zombies was not voluntary), and their sanitisation and 're-humanisation' are accomplished. Therefore, if it is true that the post-human and post-apocalyptic here thematise feminist and queer critiques of white male epistemologies, anthropocentrism, hetero-patriarchy and border regimes – as feminist scientists and post-human philosophers argue (Fox Keller 1992; Haraway 1991, 2003, 2008; Barad 2007; Alaimo and Hekman 2008; Braidotti 2013) – it is also true that the Other is described as not belonging to humanity. The post-human condition, that is, the one that sees an individual exit the human community to re-enter it as a PDS, is neither intelligible nor translatable for humans. It is through sanitisation and re-humanisation of the undead that the monster is made into a domesticated 'internal Other' and as such able/allowed to tell their story and bridge the world of the living and the afterlife. The border (Mezzadra and Sakai 2014) between the 'roaring' of the undead and human language defines the former as the language of death and thus horrific. It is only when the medicated PDS start speaking again that the horror of death can be mitigated. Language here has two important functions: it makes speaking-PDSs both accountable for what they did as zombies and responsible towards the community that reintegrated them, and it makes the trauma of resurrection a *shared* experience. Therefore, it is not empathy and compassion (which, as Hartman argues, 'obliterates' the traumatised Other, 1997, p. 19) that connect humans and post-humans, but a sort of *partage* (Iveković 2019) of the same feelings of loss, terror, hope and love. At the same time, different positionalities in the story need to be acknowledged: the story of the resurrection is not 'homolingual' (Sakai 1997) and just as in the case of multilocality and multivocality of the same language, it produces meanings that cannot and should not be reconciled into a pacified narrative.

It is the *impossibility* of fully reconciling a story (and its semantics) of violence and horrors that maintains the revenant as *risky*, even when they have been made docile by inclusion, integration and normalisation. In fact, they embody a violence (of borders, of death) that cannot be erased and which lies at the very foundation of post-apocalyptic society.

8 Conclusions

In this section, I tried to reveal the complexity of the symbolic texture shaping the imaginary that lies beneath 'our' fears of a catastrophe prompted by the loss of isolation from histories of past horrors. I have read the fears related to the mobility of the (postcolonial) Other(s) as deeply connected to repressed memories of violence and feelings of guilt on the one hand, and the 'we's' horror for a reverse colonisation on the other. Fears of a reverse colonisation are expressed through the image of the *past devouring the present* with the same ferocity with which the West devoured (and still devours, directly or indirectly) lives, stories and meanings of a distanced and immobilised Other that today is seen as imposing their proximity and presence in the former colonial metropole. I have explored existing convergences between popular film and visual news media representations of these fears, which depict Europe as a *Fortress* that must be defended from the *fast-running* and *devouring* horrors of the past, present and future, revealing the fictionality of *Fortress Europe* as the result of postcolonial anxieties. In fact, as the next section on the spatial dimension of monstrosity will show in more detail, fears of mobility have occupied Europe's space and self-perception since Christianity first and then capitalist Modernity, when the unruly and subversive colonised and enslaved were the subjects of mobility. Nevertheless, the positioned gaze of the 'we' interprets the influx of migrants and terrorism as a semiotically new entity, as if they were not part of a longer story and a broader geography. What is at stake here is rather the perception of Europe's and the West's role in history. Although Europe is not the primary destination of global migration and terrorisms, it appears so in all the visual texts I have discussed, be they news or film productions. *Fast-running* zombies attacking Europe and the West need thus to be seen as the result of a discursive operation that attempts to re-centre Europe and the West in a neocolonial discursive *geography* that has no correspondence in reality. The visual discourse of the invasion, then, is to be understood as a cultural *and* political response to that 'provincializing of Europe' (Chakrabarty 2000) that for many different reasons – geopolitical, climatic, economic, financial, social and cultural – has repositioned elsewhere, outside Europe, many of the crossroads of global power relations.

Furthermore, disquieting fantasies of annihilation assume Europe and the West as even and (culturally, socially and politically) coherent space, as if the zombie threat in film and the 'enemy' in the terror attacks that have hit Britain, Spain, France, Belgium and Germany since 2005, were not one of us and could be eradicated without undermining the social fabric of European and Western nations. Instead, what zombie films highlight, with the exception of *WWZ*, is that the threat originates inside those nations and only at a later time becomes potentially borderless and infinite. These films, and particularly *28 Days Later* and *28 Weeks Later*, thematise the idea that the *past devouring the present* (the rabid monster devouring humans) as much as *the present fighting back* (the military and civilians exterminating the threat) can be equally deadly for everybody. Therefore, not only the threat originates among us, but our response can make things worse, leading to the de-humanisation of the living.

As shown by the media and by political and academic debates on the London attacks as well as in British films that also dealt with 9/11, such as *28 Days Later* and *28 Weeks Later*, the imaginary deployed when facing the threat contemplates three options for the human community. The first option is an extreme version of *mors tua vita mea* (annihilation in *28 Weeks Later*, which recalls the 2001 war in Afghanistan and the 2003 war in Iraq; but also the shooting of zombies outside Jerusalem in *WWZ*, which recalls the Israeli bombing of Palestine).

The second option is a regime of segregation and totalitarian social control, which in *28 Days Later* recalls the 'let-them-die' camps in former colonies, WW2 concentration camps, and the necropolitics at sea.

The third option is a new 'affective community' grounded in collective atonement for past and present violence and discrimination (as exemplified by *In the Flesh*). In the last case, the memory of the colonial Holocaust is allegorised in the genocide of the post-humans, in a zombie version of Saint John's apocalypse where the sense of guilt for having exterminated the dead is partially relieved through the sanitisation of some of them. I have used this mini-series to illustrate the ambiguity of concepts like rescue, salvation, sanitisation, inclusion and integration in terms of the tension between empathy and compassion on one hand and control/obliteration of the object of those feelings on the other. In fact, all these terms retain a colonial character that, when dealing with someone needing to be rescued (from the violence of their 'brown' fellow citizens, fathers, husbands and sons, from the harm of a violent state, from a fundamentalist group, or from the threat of climate-change impacts), reproduce the unchallenged power relations that made them fleeing subjects in the first place. They, in a few words, 'contribute to an ideological formation that disconnects connected histories and that divorces the contemporary [migrants] crisis from Europe's long history of empire and racial violence' (Danewid 2017, p. 1679). This is made clear, for instance in Ida Danewid's critique of the German art group, The Center for Political Beauty, whose performance in front of the Reichtag in Berlin on 15 June 2015 was instead praised by Jill Casid:

> Take, for instance, the media-driven, grave-digging scenes performed by the Berlin-based activist theatre group, the Center for Political

> Beauty [CPB], under the headline threat of the uncanny as a force of political retribution: 'Die Toten kommen/The Dead are Coming'. This group self-described as an aesthetic 'assault team' perform in the 'as if' of coffin-carrying demonstrations and crosses planted on the lawn of the Reichstag (with phrases such as 'borders kill'). [...] They proclaim: 'The Center for Political Beauty took these dead immigrants from the EU's external borders right to the heart of Europe's mechanism of defence: to the German capital. Those who died of thirst or hunger at our borders on their way to a new life, were thus able to reach the destination of their dreams beyond their death. Together with the victims' relatives, we opened inhumane graves, identified and exhumed the bodies and brought them to Germany' (The Center for Political Beauty).
>
> (Casid 2018, p. 205)

> Pro-refugee groups such as the CPB are praised for asserting 'a politics of mourning that disrupts the script of nationalist kinship' [...] and for scandalising what makes migrant deaths possible in the first place. Grief for unknown others – for strangers – is here understood as offering a radical challenge to the xenophobia and white nationalism that underwrite the necropolitical logic of the European border regime. When viewed closely, these calls for rescue, welcome, and hospitality nonetheless confirm rather than disturb colonial relations of power. As we shall see, by seeking to extend 'grief and care to the dead stranger', [...] they contribute to an ideological formation that disconnects connected histories and that divorces the contemporary Mediterranean crisis from Europe's long history of empire and racial violence.
>
> (Danewid 2017, p. 1679)

On the other hand, the possibility of challenging those same power relations through inclusion is grounded in a shared (*partagée*) grammar of trauma as well as in the common struggle against injustice. In fact, while zombie films thematise fears of endless repetition, spatial unboundedness and the consequent erasure of *Time* and *geography* through the manhunts against rabid monsters, what is implicit is that quarantines, colonies, camps, prisons and borders cannot serve as solutions but only as temporary means. The best future envisioned is that of a new society grounded in co-existence through language and based on a radical re-conceptualisation (decolonisation) of time and space.

Notes

1 Fanon built his idea of ethno-psychiatry on the collective dimensions of trauma as a critique of Freud's psychoanalytical theory of the *uncanny* as an individual experience (1919). See also Gibson (2003), and Gibson and Beneduce (2017).

2 For a discussion of the monstrous feminine in Modern and post-Modern Western societies see, among others, Irigaray (1977); Haraway (1992); Braidotti (1997);

Creed (2000); Shildrick (2002); Federici (2004); Santos (2016). See also Section 2 of this book.

3 'who devour human flesh and walk around naked [...] engaging in the most despicable crimes of lust and sodomy' (my translation).

4 See Gerbi (2000 [1983 (1955)], 1975), Landucci (1972), Pagden (1987), Kuper (1988).

5 My interpretation of 'repressed memories' is consonant with Stoler's colonial 'aphasia' (2016) and with that which colonial historiography calls 'selective memory' (see Giuliani and Lombardi-Diop 2013 and Giuliani 2015).

6 www.bbc.com/news/uk-11663405.

7 For an interesting analysis of objectivation and simplification in the musealisation of Atlantic and Mediterranean Middle Passages, see Gatta and Muzzopapa (2012).

8 Heller Charles and Lorenzo Pezzani (Forensic Architecture Collective). *Liquid Traces: The Left-to-Die Boat Case*. UK, 2014. https://forensic-architecture.org/about/agency.

9 In *Caché* (2005), George is an affluent Parisian who lives in his sumptuous house with his wife, Anne. His memory of a childhood lie about an Algerian friend comes back to haunt him when he starts receiving videotapes, the last of which shows the brutal suicide of a man whom he discovers is his friend's son.

10 www.theguardian.com/commentisfree/2015/apr/20/lampedusa-refugee-fleeing-libya-boats-italy;www.telegraph.co.uk/news/picturegalleries/worldnews/8413997/Illegal-migrants-from-Tunisia-and-Libya-flood-the-tiny-Italian-island-of-Lampedusa.html.

11 https://news.un.org/en/story/2013/10/453012-after-latest-lampedusa-tragedy-ban-calls-action-protect-human-rights-migrants; www.dailystar.com.lb/News/World/2019/Mar-06/478143-more-than-80-migrants-rescued-off-italys-lampedusa.ashx.

12 https://africanmanager.com/migrants-pres-de-lampedusa-rome-menace-de-les-renvoyer-en-libye/.

13 www.tdg.ch/monde/europe/L-UE-se-penche-sur-sa-politique-dimmigration/story/15511242.

14 In this series, the plot is very intimate and private, dealing with the trauma of the zombie apocalypse and the traumatised body of the undead in a small Lancashire town, the fictional Roarton (evocatively named, hinting at the roaring sounds made by zombies). The series is set a few years after 'the rising', as the zombie apocalypse is called, when the government tries to rehabilitate the un-dead and reintegrate them into society. Kieren Walker returns, after being rehabilitated as a PDS (Partially Deceased Syndrome sufferer), and is welcomed by his parents who consider his return as a gift, a second chance at correcting parental mistakes made during their son's 'first life'. The rehabilitation process restores his memories with chemicals that doctors – and later family members – inject into his spine through a little hole cut below his neck. This process enables – physically and socially – the until then voiceless posthumans to talk about their traumatic journey across life and death.

References

Acosta, J. de (1577) *De procuranda Indorum salute*, re-published as *Predicación del evangelio en las Indias*. Madrid: Atlas (this edition 1954).

Agamben, G. (1995) *Homo Sacer: Sovereign Power and Bare Life*. Stanford: Stanford University Press (this edition 1998).

Agamben, G. (2003) *Stato di Eccezione. Homo sacer II*. Turin: Bollati Boringhieri.

Agier, M. (2008) *Gérer les indésirables. Des camps des réfugiés au gouvernement humanitaire*. Paris: Flammarion.

Agier, M. (2018) *The Jungle: Calais's Camps and Migrants*. Cambridge, UK: Polity Press.

Ajana, B. (2013) *Governing through Biometrics: The Biopolitics of Identity*. Basingstoke: Palgrave.

Alaimo, S. and Hekman, S. (eds) (2008) *Material Feminisms*. Bloomington: Indiana University Press.

Allen, T. W. (1994) *The Invention of the White Race*. London: Verso (this edition 2012).

Amoore, L. and De Goede, M. (eds) (2008) *Risk and the War on Terror*. New York: Routledge. https://doi.org/10.4324/9780203927700

Aradau, C. (2004) 'The perverse politics of four-letter words: risk and pity in the securitisation of human trafficking'. *Millennium: Journal of International Studies* 33 (2), pp. 251–277. https://doi.org/10.1177/03058298040330020101

Aprilia, G. A. and Soelistyarini, T. D. (2016) 'The representation of African Americans in I am Legend and World War Z'. *Unair* 5 (1), pp. 46–55. http://dx.doi.org/10.18551/rjoas.2019-01.08

Asad, T. (2007) *On Suicide Bombing*. New York: Columbia University Press.

Asad, T., Butler, J., Mahmood, S. and Brown, W. (2009) *Is Critique Secular? Blasphemy, Injury, and Free Speech*. Berkeley, Los Angeles and London: University of California Press.

Ashcroft, B. (2008) *Caliban's Voice: The Transformation of English in Post-Colonial Literatures*. London: Routledge.

Avramescu, C. (2003) *An Intellectual History of Cannibalism*. Princeton: Princeton University Press (this edition 2011).

Baishya, A. K. (2011) 'Trauma, post-apocalyptic science fiction and the post-human'. *Wide Screen* 3 (1), pp. 1–25.

Bakewell, O. (2008) '"Keeping them in their place": the ambivalent relationship between development and migration in Africa'. *Third World Quarterly* 29 (7), pp. 1341–1358. http://doi.org/10.1080/01436590802386492

Baldwin, A. (2013) 'Racialisation and the figure of the climate-change migrant'. *Environment and Planning A* 45, pp. 1474–1490. https://doi.org/10.1068/a45388

Banivanua-Mar, T. (2005) 'Consolidating violence and colonial rule. Discipline and protection in colonial Queensland'. *Postcolonial Studies* 8 (3), pp. 303–319. https://doi.org/10.1080/13688790500231053

Banivanua-Mar, T. (2007) *Violence and Colonial Dialogue: The Australian-Pacific Indentured Labour Trade*. Honolulu: University of Hawaii Press.

Barad, K. (2007) *Meeting the Universe Halfway: Quantum Physics and the Entanglement of Matter and Meaning*. Durham: Duke University Press. http://doi.org/10.1215/9780822388128

Barker, F., Hulme, P. and Iversen, M. (eds) (1998) *Cannibalism and the Colonial World*. Cambridge, UK: Cambridge University Press.

Bartolovich, C. (1988) 'Consumerism, or the cultural logic of late cannibalism'. In: Barker, F., Hulme, P. and Iversen, M. (1988) *Cannibalism and the Colonial World*, Cambridge, UK: Cambridge University Press, pp. 204–237.

Baudrillard, J. (2005) *The Conspiracy of Art. Manifestos, Interviews, Essays*. New York and Los Angeles: Semiotext(e).

Benjamin, W. (1940) 'Theses on the philosophy of history'. In: Benjamin, W. *Illuminations: Essays and Reflections*. New York: Schocken Books, pp. 253–264 (this edition 1968).

Bentham, J. (1768) *Essay on the First Principles of Government*. London: Routledge (this edition 2011).

Berglund, J. (2006) *Cannibal Fictions: American Explorations of Colonialism, Race, Gender and Sexuality*. Madison and London: University of Wisconsin Press.

Bettini, G. (2013) 'Climate barbarians at the gate? A critique of apocalyptic narratives on "climate refugees"'. *Geoforum* 45, pp. 63–72. https://doi.org/10.1016/j.geoforum.2012.09.009

Biber, K. (2005) 'Cannibalism and colonialism'. *Sydney Law Review* 27 (4), pp. 623–637.

Bigo, D. (2002) 'Security and immigration: toward a critique of the governmentality of unease'. *Alternatives: Global, Local, Political* 27 (1), pp. 63–92. https://doi.org/10.1177/03043754020270S105

Bishop, K. W. (2010) *American Zombie Gothic*. Jefferson: McFarland & Co.

Blanchard, P., Bancel, N., Deroo, E. and Lemaire, S. (2002) *Human Zoos: Science and Spectacle in the Age of Colonial Empires*. Liverpool: Liverpool University Press (this edition 2008).

Blanchard, P., Bancel, N. and Lemaire, S. (2005) *La fracture coloniale: la société française au prisme de l'héritage colonial*. Paris: Gallimard.

Braidotti, R. (1997) 'Mothers, monsters, and machines'. In: Conboy, K. and Medina, N. (eds) *Writing on the Body: Female Embodiment and Feminist Theory*. New York: Columbia University Press, pp. 59–79.

Braidotti, R. (2013) *Post-human*. Cambridge, UK: Polity Press.

Brioni, S. (2013) 'Costruire la razza come mostruosità in Italia'. In: Giuliani, G. (ed.) 'La sottile linea bianca'. Special issue. *Studi Culturali* 10 (1), pp. 293–303.

Broderick, M. (1993) 'Surviving Armageddon: beyond the imagination of disaster'. *Science Fiction* 61 (20), pp. 342–382.

Browning, J. and Picart, C. J. K. (2009) *Draculas, Vampires, and Other Undead Forms. Es-says on Gender, Race, and Culture*. Lanham, Toronto, New York, Plymouth: The Scarecrow Press.

Butler, J. (1993) 'Engendered/endangered: schematic racism and white paranoia'. In: Gooding-Williams, R. (ed.) *Reading Rodney King, Reading Urban Uprising*. New York and London: Routledge, pp. 15–22.

Butler, J. (2008) 'Sexual politics, torture, and secular time'. *British Journal of Sociology* 59 (1), pp. 1–23. https://doi.org/10.1111/j.1468-4446.2007.00176.x

Butler, J. (2009) *Frames of War. When Is Life Grievable?* New York: Verso.

Buzan, B., Wæver, O. and de Wilde, J. (1998) *Security: A New Framework for Analysis*, Boulder: Lynne Rienner.

Calder Williams, E. (2010) *Combined and Uneven Apocalypse: Luciferian Marxism*. Winchester and Washington: Zero Books.

Canetti, E. (1981) *Crowds and Power*. New York: The Viking Press.

Casid, J. H. (2018) 'Necropolitics at sea'. In: Gualtieri, C. (ed.) *Migration and the Contemporary Mediterranean. Shifting Cultures in Twenty-First Century Italy and Beyond*. Oxford, Bern, Berlin, Brussels, New York and Wien: Peter Lang, pp. 193–214.

Castles, S. (2010) 'Understanding global migration: a social transformation perspective'. *Journal of Ethnic and Migration Studies* 36 (10), pp. 1565–1586. https://doi.org/10.1080/1369183X.2010.489381

Chakrabarty, D. (2000) *Provincialising Europe*. Princeton: Princeton University Press.

Christie, D. and Lauro, S. J. (eds) (2011) *Better off Dead: The Evolution of the Zombie as Post-human*, New York: Fordham University Press.

Conrad, J. (1899) *Heart of Darkness*. New York: Tribeca Books (this edition 2010).

Coulombe, M. (2012) *Petite philosophie du zombie*. Paris: PUF.

Creed, B. (2000) 'Alien and the monstrous-feminine'. In: Kirkup, G., Hovenden, F., Janes, L. and Woodward, K. (eds) *The Gendered Cyborg: A Reader*. London and New York: Routledge, pp. 110–135.

Creed, B. and Hoorn, J. (eds) (2002) *Body Trade: Captivity, Cannibalism and Colonialism in the Pacific*. New York: Routledge.

Cunningham, D. and Warwick, A. (2018) 'The ambassadors of nil: notes on the zombie apocalypse'. In: Germanà, M. and Mousoutzanis, A. (eds) *Apocalyptic Discourse in Contemporary Culture*. London: Routledge, pp. 175–192. https://doi.org/10.4324/9781315883861

Danewid, I. (2017) 'White innocence in the Black Mediterranean: hospitality and the erasure of history'. *Third World Quarterly* 38 (7), pp. 1674–1689. https://doi.org/10.1080/01436597.2017.1331123

Dante, *Inferno*, trans. J. D. Sinclair (1961) London: Oxford University Press.

Davis, H. and Todd, Z. (2017) 'On the importance of a date, or decolonizing the Anthropocene'. *ACME: An International Journal for Critical Geographies* 16 (4), pp. 761–780.

Davis, J., Moulton, A. A., Van Sant, L. and Williams, B. (2019) 'Anthropocene, capitalocene, … plantationocene?: a manifesto for ecological justice in an age of global crises'. *Geography Compass* 13 (5), e12438. https://doi.org/10.1111/gec3.12438

De Certeau, M. (1975) *The Writing of History*. New York: Columbia University Press (this edition 1988).

De Genova, N. and Peutz, N. (eds) (2010) *The Deportation Regime: Sovereignty, Space and the Freedom of Movement*. Durham: Duke University Press. https://doi.org/10.1215/9780822391340

De las Casas, B. (1565) *A Short Account of the Destruction of the Indies*. London: Penguin Classics (this edition 1992).

De Paw, C. (1771) *Recherches philosophiques sur les Américains, ou Mémoires intéressants pour servir à l'Histoire de l'Espèce Humaine. Avec une Dissertation sur l'Amérique & les Américains*. Charleston: Nabu Press (this edition 2012).

Del Lucchese, F. (2014) 'When the slaves go marching out: indignatio, invisible bodies, and political theory'. *Citizenship Studies*, 18 (5), pp. 549–561. https://doi.org/10.1080/13621025.2014.923706

Del Lucchese, F. (2019) *Monstrosity and Philosophy: Radical Otherness in Greek and Latin Culture*. Edinburgh: Edinburgh University Press.

Deltombe, T. and Rigouste, M. (2005) 'L'ennemi intérieur: la construction médiatique de la figure de l'"Arabe"'. In: Blanchard, P., Bancel, N. and Lemaire, S. (eds) *La fracture coloniale*, pp. 195–202.

Dendle, P. (2012) *The Zombie Movie Encyclopedia, volume 2: 2000–2010*. Jefferson: McFarland & Company.

Douglass, F. (1855) *My Bondage and My Freedom*. New York: Penguin (this edition 2003).

Douglass, F. (1881) *Life and Times of Frederick Douglass*. New York: Dover Publications (this edition 2003).

Esposito, R. (2008) *Bíos: Biopolitics and Philosophy*. Minneapolis: University of Minnesota Press.

Fanon, F. (1952) *Black Skin, White Masks*. New York: Grove Press (this edition 1994).

Fassin, D. (2011) *Humanitarian Reason*. Berkeley, Los Angeles and London: University of California Press.

Federici, S. (2004) *Caliban and the Witch: Women, the Body and Primitive Accumulation*. Brooklyn NY: Autonomedia.

Fekete, L. (2009) *A Suitable Enemy. Racism, Migration and Islamophobia in Europe*. New York, Pluto Press.

Foucault, M. (1961) *History of Madness*. London and New York: Routledge (this edition 2006).

Foucault, M. (1976) *Society Must Be Defended*. New York: Picador (this edition 2003).

Freud, S. (1919) *The Uncanny*. New York: Penguin (this edition 2003).

Friedlander, H. (1997) *The Origins of Nazi Genocide. From Euthanasia to the Final Solution*. Chapel Hill: University of North Carolina Press.

Furedi, F. (2007) *Invitation to Terror. The Expanding Empire of the Unknown*. London: Continuum.

Gatta, G. and Muzzopapa, G. (2012) '"Middle passages", musealizzazione e soggettività a Bristol e Lampedusa in Estetica'. *Studi e Ricerche* 1, pp. 167–181. http://doi.org/10.4434/ESR.20396635.032012.11

Gerbi, A. (1955) *La disputa del Nuovo Mondo. Storia di una polemica, 1750–1900*. Milan: Adelphi (this edition 2000).

Gerbi, A. (1975) *La natura delle Indie Nove. Da Cristoforo Colombo a Gonzalo Fernández de Oviedo*. Milan-Naples: Ricciardi.

Gibson, N. (2003) *Fanon: The Postcolonial Imagination*. Cambridge, UK: Polity Press.

Gibson, N. and Beneduce, R. (2017) *Frantz Fanon, Psychiatry and Politics*. London: Rowman & Littlefield.

Gilroy, P. (1993) *The Black Atlantic: Modernity and Double-Consciousness*. New York: Verso.

Gilroy, P. (2000) *Against Race. Imagining Political Culture beyond the Color Line*. Cambridge, MA: Harvard University Press.

Gilroy, P. (2002) *Between Camps: Nations, Cultures and the Allure of Race: Nations, Culture and the Allure of Race*. London: Routledge.

Giuliani, G. (2006) 'Il concetto di impero nel pensiero politico inglese tra il XVII e la prima metà del XIX secolo'. *Il Pensiero politico* 1, pp. 3–35.

Giuliani, G. and Lombardi-Diop, C (2013) 'Introduzione'. In Giuliani, G. and Lombardi-Diop, C. (2013) *Bianco e nero. Storia dell'identità razziale degli italiani*. Florence-Milan: Le Monnier/Mondadori Education, pp. 1-20.

Giuliani, G. (2015) 'Introduzione'. In Giuliani, G. (ed.) *Il colore della nazione*. Florence-Milan: Le Monnier/Mondadori Education, pp. 1–16.

Giuliani, G. (2016) *Zombie alieni e mutanti. Le Paure dall'11 settembre a oggi*. Florence–Milan: Le Monnier/Mondadori Education.

Giuliani, G. (2019) *Race, Nation and Gender in Modern Italy. Intersectional Representations in Visual Culture*. London: Palgrave Macmillan. http://doi.org/10.1057/978-1-137-50917-8

Giuliani, G. (2020 forthcoming) 'Monstrous beauties: bodies in motion between colonial archives and the migrant and refugee crisis'. In: Tate, S. A. (ed.) *The Palgrave Handbook on Critical Race and Gender*. London: Palgrave.

Giuliani, G. and Vacchiano, F. (2019) 'Il senso degli italiani (e degli europei) per l'Altrx che vien dal mare. Cinema, bianchezza e Middle passage'. In: de Franceschi, L. and Polato, F. (eds) *Narrazioni postcoloniali della contemporaneità, tra conflitto e convivenza/Postcolonial Narratives of the Present, between Conflict and Coexistence*. Special issue *Imago* 10 (19), pp. 137–152.

Graham, E. L. (2002) *Representations of the Post/Human. Monsters, Aliens and Others in Popular Culture*. New Brunswick: Rutgers.
Grebowicz, M. (ed.) (2007) *SciFi in the Mind's Eye. Reading Science through Science Fiction*. Chicago: Open Court.
Haas, R. (2008) 'The age of nonpolarity: what will follow U.S. dominance?' *Foreign Affairs* 87 (3), pp. 44–56.
Hage, G. (2016) 'État de siège: A dying domesticating colonialism?' *American Ethnologist* 43 (1), pp. 38–49. http://doi.org/10.1111/amet.12261
Haraway, D. J. (1991) 'A cyborg manifesto: science, technology, and socialist-feminism in the late twentieth century'. In: Haraway, D. J. *Simians, Cyborgs and Women: The Reinvention of Nature*. New York: Routledge, and London: Free Association Books.
Haraway, D. J. (1992) 'Ecce homo, ain't (ar'n't) I a woman, and inappropriate/d others: the human in a post-humanist landscape'. In: Butler, J. and Scott, J. W. (eds) *Feminists Theorise the Political*. London: Routledge, pp. 86–100.
Haraway, D. J. (2003) *The Companion Species Manifesto: Dogs, People, and Significant Otherness*. Chicago: Prickly Paradigm Press.
Haraway, D. J. (2008) *When Species Meet*. Minneapolis: University of Minnesota Press.
Harrington, J. (1656) *Oceana*. In: *The Commonwealth of Oceana and A System of Politics*. Cambridge, UK: Cambridge University Press (this edition 1992).
Hartman, S. V. (1997) *Scenes of Subjection: Terror, Slavery, and Self-Making in Nineteenth-Century America*. Oxford: Oxford University Press.
Hobbes, T. (1651) *Leviathan*. London: Penguin Classics (this edition 2017).
hooks, b. (1981) *Ain't I a Woman?* Boston: South End Press.
Hurley, K. (1996) *The Gothic Body: Sexuality, Materialism, and Degeneration at the Fin de Siecle*. Cambridge: Cambridge University Press.
Huysmans, J. (2006) *The Politics of Insecurity: Fear, Migration and Asylum in the EU*. London: Routledge. http://doi.org/10.4324/9780203008690
Irigaray, L. (1977) *This Sex Which Is Not One*. Ithaca: Cornell (this edition 2008).
Iveković, R. (2019) *Politiques de la traduction. Exercices de partage*. Marseille: TERRA-HN éditions Alterego.
James, C. L. R. (1938) *The Black Jacobins: Toussaint L'Ouverture and the San Domingo Revolution*. New York: Vintage Books (this edition 1989).
Jáuregui, C. A. (2008) *Canibalia. Canibalismo, calibanismo, antropofagia cultural y consumo en América Latina*. Madrid: Iberoamericana.
Jensen, L. (2018) *Postcolonial Denmark. Nation Narration in a Crisis Ridden Denmark*. Abingdon and New York: Routledge.
Jensen, L. and Loftsdóttir, K. (eds) (2012) *Whiteness and Postcolonialism in the Nordic Region. Exceptionalism, Migrant Others and National Identities*. Farnham and Burlington: Ashgate.
Jones, D. J. (2014) *Sexuality and the Gothic Magic Lantern: Desire, Eroticism and Literary Visibilities from Byron to Bram Stoker*. Basingstoke: Palgrave Macmillan. http://doi.org/10.1057/9781137298928
Jones, R. (2016) *Violent Borders: Refugees and the Right to Move*. [ebook reader] London and New York: Verso.
Kee, C. (2011) '"They are not men … they are dead bodies". From Cannibal to Zombie and back again'. In: Christie, D. and Lauro, S. J. (eds) *Better off Dead*. New York: Fordham University Press, pp. 9–23.

Keller, E. F. (1992) *Inventing Women: Science, Technology and Gender*. Cambridge, UK: Polity Press.

Khosravi, S. (2010) *Illegal Traveller: An Auto-Ethnography of Borders*. Basingstoke: Palgrave.

Khosravi, S. (ed.) (2018) *After Deportation: Ethnographic Perspectives*. Basingstoke: Palgrave.

Kirkup, G., Janes, L., Hovenden, F. and Woodward, K. (eds) (2000) *The Gendered Cyborg: A Reader*. London and New York: Routledge.

Koselleck, R. (1979) *Futures Past. On the Semantics of Historical Time*. New York: Columbia University Press (this edition 2004).

Kuper, A. (1988) *The Invention of Primitive Society: Transformations of an Illusion*. London: Routledge.

Lacan, J. (1973) *The Four Fundamental Concepts of Psycho-Analysis*. London: Penguin (this edition 1994).

Lafitau, J. F. (1724) *Mœurs des sauvages américains comparées aux mœurs des premiers temps*. Charleston: Nabu Press (this edition 2011).

Landucci, S. (1972) *I filosofi e i selvaggi*. Turin: Einaudi.

Langer, P. P. (2010) '"Piores que bestas feras": Garcilaso de la Vega e o imaginário hispano-inca sobre os Guarani Chiriguano' *Topoi* 11 (21) http://dx.doi.org/10.1590/2237-101X011021001

Lindqvist, S. (1997) *Exterminate All The Brutes*. New York: The New Press.

Lowenstein, A. (2005) *Shocking Representations. Historical Trauma, National Cinema, and the Modern Horror Film*. New York: Columbia University Press.

Lucci, A. (2014) 'Homo homini zombie. Su morale e stato di natura, da Thomas Hobbes a *The Walking Dead*'. *Popsophia/Lo Sguardo-rivista di filosofia* 16 (3), pp. 109–115.

Luckhurst, R. J. (2012) *The Mummy's Curse: The True History of a Dark Fantasy*. Oxford: Oxford University Press.

Luckhurst, R. J. (2015) *Zombies. A Cultural History*. London: Reaktion Books.

Locke, J. (1690) *Second Treatise on Government*. In: *Two Treatises of Civil Government*. London: Dent and New York: Duttond & Co. (this edition 1962).

Matheson, R. B. (1954) *I am Legend*. New York: Gold Medal Books.

Mazzara, F. (2018) *Reframing Migration: Lampedusa, Border Spectacle and the Aesthetics of Subversion*. Oxford and New York: Peter Lang.

Mbembe, A. (2003) 'Necropolitics'. *Public Culture* 15 (1), pp. 11–40. https://doi.org/10.1215/08992363-15-1-11

McClintock, A. (1995) *Imperial Leather: Race, Gender and Sexuality in the Imperial Context*. London and New York: Routledge.

McClintock, A. (2009) 'Paranoid Empire: Specters from Guantánamo and Abu Ghraib'. *Small Axe* 13 (1), pp. 50–74. http://doi.org/10.1215/07990537-2008-006

Mezzadra, S. (2001) *Diritto di fuga. Migrazioni, cittadinanza, globalizzazione*. Verona: Ombre corte.

Mezzadra, S. (2004) 'The right to escape'. *ephemera* 4 (3), pp. 267–285.

Mezzadra, S. (2015) 'The proliferation of borders and the right to escape'. In: Jansen, Y., Celikates, R. and de Bloois, J. (eds) *The Irregularization of Migration in Contemporary Europe. Detention, Deportation, Drowning*. London and New York: Rowman & Littlefield, pp. 121–135.

Mezzadra, S. and Rahola, F. (2015) 'The postcolonial condition: a few notes on the quality of historical time in the global Present'. In: Mareddy, P. K., Heidemann, B. and Laursen, O. B. (eds) (2015) *Reworking Postcolonialism. Globalization, Labour and Rights*. London: Palgrave Macmillan, pp. 36–54

Mezzadra, S. and Neilson, B. (2013) *Border as Method, or, the Multiplication of Labor*. Durham, NC: Duke University Press.

Mezzadra, S. and Sakai, N. (2014) 'Introduction'. Special issue 'Politics', *Translation. A transdisciplinary Journal* 4, pp. 9–29.

Mirzoeff, N. (2011) *The Right to Look. A Counterhistory of Visuality*. Durham: Duke University Press.

Morrison, T. (1987) *Beloved*. New York: Vintage Books (this edition 2008).

Moulier-Boutang, Y. (1998) *De l'esclavage au salariat: Economie historique du salariat bridé*. Paris: Presses Universitaires de France.

Negri, A. and Hardt, M. (2000) *Empire*. Cambridge, MA: Harvard University Press.

Neocleous, M. (2005) *The Monstrous and the Dead: Burke, Marx, Fascism*. Chicago: University of Chicago Press.

Nì Fhlainn, S. (2011) '"All dark inside": dehumanisation and zombification in post-modern cinema'. In: Christie, D. and Lauro, S. J. (eds) *Better Off Dead*, pp. 139–157.

Nuzzo, L. (2013) 'Foucault and the enigma of the monster'. *International Journal of Semiotic Law* 26 (1), pp. 55–72. https://doi.org/10.1007/s11196-012-9275-8

Nyers, P. (2015) 'Migrant citizenships and autonomous mobilities'. *Migration, Mobility, & Displacement* 1 (1), pp. 23–39. https://doi.org/10.18357/mmd11201513521

Obeyesekere, G. (1992) '"British cannibals": contemplation of an event in the death and resurrection of James Cook, explorer'. *Critical Inquiry* 1, pp. 630–654.

Obeyesekere, G. (2001) 'Narratives of the self: Chevalier Peter Dillon's Fijian cannibal adventures'. In: Creed, B. and Hoorn, J. (eds) *Body Trade, Captivity, Cannibalism and Colonialism in the Pacific*. New York: Routledge, pp. 69–111. http://doi.org/10.1525/california/9780520243071.003.0007

Pagden, A. (1987) *The Fall of Natural Man: The American Indian and the Origins of Comparative Ethnology*. Cambridge, UK: Cambridge University Press.

Papadopoulos, D. and Tsianos, V. (2013) 'After citizenship: autonomy of migration, organisational ontology and mobile commons'. *Citizenship Studies* 17, pp. 178–196. https://doi.org/10.1080/13621025.2013.780736

Phillips, J. (1998) 'Cannibalism qua capitalism: the metaphorics of accumulation in Marx, Conrad, Shakespeare and Marlowe'. In: Barker, F., Hulme, P. and Iversen, M. (eds) *Cannibalism and the Colonial World*. Cambridge, UK: Cambridge University Press, 183–203.

Ponzanesi, S. and Colpani, G. (2016) *Postcolonial Transitions in Europe. Contexts, Practices and Politics*. London: Rowman & Littlefield.

Puwar, N. (2004) *Space Invaders: Race, Gender and Bodies Out of Place*. New York: Berg.

Rediker, M. (2007) *The Slave Ship: A Human History*. New York: Penguin.

Sakai, N. (1997) *Translation and Subjectivity. On Japan and Cultural Nationalism*. Minneapolis: Minnesota University Press.

Said, E. (1978) *Orientalism*. London: Routledge and Kegan Paul.

Said, E. (1993) *Culture and Imperialism*. New York: Vintage.

Santos, C. (2016) *Unbecoming Female Monsters: Witches, Vampires and Virgins*. Lanham: KY: Lexington Press.

Saucier, P. K. and Woods, T. P. (2014) 'Ex aqua: the Mediterranean Basin, Africans on the move and the politics of policing'. *Theoria: The Journal of Social and Political Theory* 141 (December), pp. 55–75. https://doi.org/10.3167/th.2014.6114104

Scarry, E. (1985) *The Body in Pain. The Making and Unmaking of the World*. Oxford: Oxford University Press.

Seed, D. (2000) *Imagining Apocalypse. Studies in Cultural Crisis*. London: Macmillan.
Sharpe, C. (2009) *Monstrous Intimacies: Making the Post-Slavery Subject*. Durham: Duke University Press. https://doi.org/10.1215/9780822391524
Sharpe, C. (2016) *In the Wake: On Blackness and Being*. Durham: Duke University Press. https://doi.org/10.1215/9780822373452
Shildrick, M. (2002) *Embodying the Monster: Encounters with the Vulnerable Self*. London: Sage. http://dx.doi.org/10.4135/9781446220573
Sloterdijk, P. (2005) *In the World Interior of Capital. For a Philosophical Theory of Globalization*. Cambridge, UK: Polity Press (this edition 2013).
Sobchack, V. (2000) *Postfuturism*. In: Kirkup, G., Hovenden, F., Janes, L. and Woodward, K. (eds) *The Gendered Cyborg*, pp. 137–147.
Sontag, S. (1965) 'The imagination of disaster'. In: Ralph, J. A. (ed.) *Hal in the Classroom: Science Fiction Film*. Dayton: Pflaum (this edition 1974), pp. 22–38.
Sousa Santos, B. (2014) *Epistemologies of the South. Justice against Epistemicide*. London: Paradigm. https://doi.org/10.4324/9781315634876
Sousa Santos, B. (2018) *The End of the Cognitive Empire. The Coming of Age of Epistemologies of the South*. Durham: Duke University Press. https://doi.org/10.1215/9781478002000
Stoler, A. L. (2016) *Duress. Imperial Durabilities in Our Times*. Durham: Duke University Press. https://doi.org/10.1215/9780822373612
Tagg, J. (1988) *The Burden of Representation: Essays on Photographies and Histories*. Amherst: University of Massachusetts Press.
Tazzioli, M. (2015) *Spaces of Governmentality. Autonomous Migration and the Arab Uprisings*. London: Rowman & Littlefield.
Tazzioli, M. (2017a) 'The government of migrant mobs. Temporary divisible multiplicities in border-zones'. *European Journal of Social Theory* 20 (4), pp. 473–490. https://doi.org/10.1177/1368431016658894
Tazzioli, M. (2017) 'Containment through mobility: migrants' spatial disobediences and the reshaping of control through the hotspot system'. *Journal of Ethnic and Migration Studies* 44 (16), pp. 2764–2779. https://doi.org/10.1080/1369183X.2017.1401514
Tazzioli, M. and Walters, W. (2016) 'The sight of migration: governmentality, visibility and Europe's contested borders'. *Global Society* 30 (3), pp. 445–464. https://doi.org/10.1080/13600826.2016.1173018
Tazzioli, M. and Garelli, G. (2018) 'Containment beyond detention: the hotspot system and disrupted migration movements across Europe'. *Environment and Planning D: Society and Space*. https://doi.org/10.1177/0263775818759335
Tazzioli, M. and Walters, W. (2019) 'Migration, solidarity and the limits of Europe'. *Global Discourse* 9 (1), pp. 175–190. https://doi.org/10.1332/204378918X15453934506030
Teti, V. (1994) *Il vampiro e la melanconia. Miti, storie, immaginazioni*. Rome: Donzelli (this edition 2018).
Ticktin, M. I. (2016) 'Thinking beyond humanitarian borders'. *Social Research: An International Quarterly* 83 (2), pp. 255–271.
Towlson, J. (2012) 'Rehabilitating Daddy. Or how horror movies say it's OK to trust authority'. *paracinema.net* Available at: http://paracinema.net/wp-content/uploads/2012/08/Rehabilitating-Daddy.pdf. Accessed 24 November 2019.
Webb, J. and Byrnand, S. (2008) 'Some kind of virus: the zombie as body and as trope'. *Body & Society* 14 (2), pp. 83–98. https://doi.org/10.1177/1357034X08090699

Williams, L. (1991) 'Film bodies: gender, genre, and excess'. *Film Quarterly* 44 (4), pp. 2–13. http://doi.org/10.2307/1212758
Wekker, G. (2016) *White Innocence. Paradoxes of Colonialism and Race*. Durham: Duke University Press.
Wells, H. G. (1898) *War of the Worlds*. Manor, TX: Arc Manor (this edition 2008).
Yusoff, K. (2018) *A Billion Black Anthropocenes or None*. [ebook reader] Minneapolis: Minnesota University press.
Žižek, S. (2008) *For They Know Not What They Do: Enjoyment as a Political Factor*. London: Verso.

Films

Balagueró, Jaume and Paco Plaza. *REC*. Spain, 2007.
Bayona, Juan A. *The Impossible*. USA and Spain, 2012.
Boyle, Danny. *28 Days Later*. UK and USA, 2002.
Craven, Wes. *Scream*. USA, 1996.
Carpenter, John. *District 13*. USA, 1976.
Carpenter, John. *Halloween*. USA, 1978.
Coppola, Francis F. *Apocalypse Now*. USA, 1979.
Dahan, Yannick and Benjamin Rocher. *The Horde*. France, 2009.
Deodato, Ruggero. *Cannibal Holocaust*. Italy, 1981.
Forster, Mark. *World War Z*. USA and Malta, 2013.
Fresnadillo, Juan C. *28 Weeks Later*. UK and Spain, 2007.
Halperin, Victor. *White Zombie*. USA, 1932.
Heller Charles and Lorenzo Pezzani (Forensic Architecture Collective). *Liquid Traces: The Left-to-Die Boat Case*. UK, 2014.
Michael Hanneke. *Caché*. France, Austria, Germany, Italy, 2005.
Hooper, Tope. *The Texas Chain Saw Massacre*. USA, 1974.
Kubrick, Stanley. *A Clockwork Orange*. UK, 1971.
Kubrick, Stanley. *Full Metal Jacket*. USA, 1987.
Lawrence, Francis. *I am Legend*. USA, 2007.
Meirelles, Fernando (2008), *Blindness*. USA and Brazil, 2008.
Nelson, Ralph. *Soldier Blue*. USA, 1970.
Ragona, Ubaldo and Sidney Salkow. *The Last Man on Earth*. Italy and USA, 1964.
Romero, George. *Night of the Living Dead*. USA, 1968.
Romero, George. *Dawn of the Dead*. USA, 1978.
Romero, George. *Land of the Dead*. USA, 2006.
Rosi, Francesco. *Fuocoammare*. Italy, 2016.
Sagal, Boris. *The Omega Man*. USA, 1971.
Spielberg, Steven. *E.T.* USA, 1982.
Terwilliger, George. *Ouanga*. USA, 1936.
Tourneur, Jacques. *I Walked with a Zombie*. USA, 1943.
Yarbrough, Jean. *King of the Zombies*. USA, 1941.

TV Series

Darabont, Frank. *The Walking Dead*. USA, 2010– (ten seasons).
Mitchell, Dominic and Jonny Campbell. *In the Flesh*. UK, 2013–2015 (two seasons).

2 *Alien-ing* the migrant

On Anthropocenic geographies of monstrosity

1 Introduction

In his 1908 groundbreaking reflection on the 'stranger', the German sociologist Georg Simmel differentiates between the stranger 'who comes today and stays tomorrow' and the extraterrestrial alien:

> The inhabitants of the star Sirius are not actually strangers to us – at least not in the sense of the word that comes into sociological consideration – but they do not exist at all for us, they stand outside of far and near. The stranger is a member of the group itself, not different from the poor and the various 'inner enemies' – an element whose immanent presence and membership include at the same time an externality and opposition. Now the pattern wherein repelling and distancing moments here comprise a form of togetherness and interacting unity may be outlined [...].
>
> (2009, p. 601)

What if Sirius' inhabitants were not in the least 'outside of far and near', and what if they knocked on our doors? What if, rather than an unlikely event, their mobility towards the 'we's' world were an allegory for an *ontologically distanced* Other coming from a real or fictional geographically and politically severed place perceived as remote and scary? Just as the stranger is not really a stranger, so Siriusians would no longer be extraterrestrials if we looked at borders from a different perspective and acknowledged that 'as far as nature is concerned, every boundary placement is arbitrary, even in the case of an insular [or planetary] situation' (Simmel 2009, p. 549). What if other-worldliness was just another symbolic *dispositif* to project strangeness and fears onto someone who (or something that) is not really an alien? In *Excursus on Social Boundary*, Simmel wrote in 1903 that:

> The form of spatial nearness or distance does not generate the peculiar phenomena of neighborliness or alienation, however inevitably it may seem. Rather even these are facts generated purely by psychological

> contents, the course of which stands in relationship to its spatial form in principle no differently than a battle or a telephone conversation to that of theirs – thus doubtlessly these processes too can be realized then only under quite specific spatial conditions. Not space, but the psychologically consequential organization and concentration of its parts have social significance.
>
> (2009, p. 544)

Unlike Simmel's analysis, in my reading of other-worldliness the stranger and the alien converge in their common *alien-ing*. Theirs is a dimension that has been *alien-ed/alienised* by virtue of a conception of space that, integrating Butler's (2009) reflections on the dehumanisation of the enemy into the notion of *geontopower* (Povinelli 2016), may be described as separating Life (human and non-human) from Nonlife (the world Out there) while assigning to non-human or lesser-human Life an ontology of monstrosity and expendability. In this conception, an 'abyssal line' (Sousa Santos 2014, pp. 118–135) is drawn that projects onto what or whom has been severed from the space of the 'we' the fear of a subversion of the spatial Order defining the human *polis*.

As I have argued in the introduction to this book, the creation of a we/they dichotomy has underpinned Western political thought since Aristotle. The idea that 'we' and 'they' are spatially, ideologically and even biophysically opposite, which also provides a rationale for the differential access to privilege and power, has been foundational to notions of *polis*, society, state and international relations for centuries. *Politics*, according to Western political philosophers, is the arena in which a conflict between enemies or its negotiation take place (Portinaro 1992). Since 9/11 this opposition has extended to supposedly distant 'civilisations' whose identities, as they appear today, have been constructed discursively since Modernity. The issue at stake here, the use of essentialising categories as paired opposites, is not new. As I argued in the first section, the practice of animalising and criminalising the Other and turning them into the enemy of civilisation has colonial roots. It exceeds the distinction between *inimicus* and *hostis* drawn by Greek philosophers and rearticulated across time, the binary of friend/enemy hypostatised by Carl Schmitt in *The Concept of the Political* (1932), and even post-WWII conceptions of 'totalitarian enemy' – the 'terrorist' described by Schmitt in *Nomos of the Earth* (1950). As I wrote earlier, it is more analogous to the embodied and spatialised concept of monstrified 'expendable subjects' highlighted by Mbembe (2003), Asad (2007) and Butler (2009) and never investigated in traditional Western political philosophy, which has always regarded colonialism and its genocides as outside the realm of *politics*.

The present section focuses on *geography*, meaning the space (the *planet* as the environment and the *world* as the human habitat) in which political operations of *bordering*, *Outsiding* and *Other(world)ing* are performed. The starting point for my reflections here is that human and non-human entities (including the environment and several of its beings and phenomena) are generally labelled as *alien* in Western imagination when they embody a transgression of institutionally sanctioned cosmogonies and taxonomies or are perceived as not fitting in, or when their mobility is troubling (Neocleous 2005, p. 28). Therefore, they are not in and of themselves ontologically distant, threatening or monstrous: their construction as such depends instead on who is exercising the 'right to look' and classify (Mirzoeff 2011), on the analytical tools of classification, and on the underlying values.

From this perspective, just like rabid zombies the inhabitants of Sirius are seemingly much less distant than in Simmel's argument. Their being 'outside of far and near' is seen here as a result of an *original border* between order and disorder, the known and the unknown, the reassuring and the frightening.

> Monsters serve both to mark the fault-lines but also, subversively, to signal the fragility of [...] boundaries [between humans and almost-humans]. They are truly monstrous [...] in their simultaneous demonstration and destabilization of the demarcations by which cultures have separated nature from artifice, human from non-human, normal from pathological [...]. In their capacity to show up the leakiness of bodily boundaries [...] this emergent array of hybrid creatures are arguably 'monstrous' not so much in the horror they evoke but in their exposure of the redundancy and instability of the ontological hygiene of the human subject.
>
> (Graham 2002, p. 12)

Nonetheless, the leaking boundaries mentioned by Graham need not be exclusively bodily but also include those set by political and social institutions and 'civilisations' identified as European and Western in contrast to the 'Out there'. The environmental, social and political history of humans and the Earth is the result of a segmentation along cartographic lines that throughout the Anthropocene have been drawn, to paraphrase De Certeau (1975), for the purpose of absorbing and eliminating biological and cultural diversity and hybridity through *geography* and *history* – combined with the subordination of all to the logic of profit (Dussel 1995; Svampa 2019; Cole 2000; Clifford 1988; Breidenbach and Zukrigl 2000; Povinelli 2016). In Western Modernity this allowed the State to build and defend its sovereignty (Benton 2010; Scott 1998). From this perspective, *bordering*, *Outsiding* and *Other(world)ing* end up being biopolitical *dispositifs* to separate people and ward off some while linking them to monstrosity, as well as distancing their geographies and projecting them onto otherworldliness. Historical examples of *Othering/Other(world)ing* can be seen in the relations between Southern Europe and Africa/the Middle East (Chambers 2008), Australia and Southern Asia (Perera 2009), North America and Latin America (Vila 2003), where practices of *alien-ing* have implied cultural separateness and racial incompatibility, although cultural and biological mixing have always been the result of more or less violent encounters, exchanges and conquests. In all these cases, languages, histories and populations have since been considered extraterritorial, foreign, alien to Western experience when they are, in fact, connected. As I have argued in the introduction when discussing the distinction between catastrophe and disaster and between places *of* disaster and places *for* disaster, the biopolitics of *bordering*, *Outsiding* and *Other(world)ing* has enforced not only a human ontology of distance and opposition through camps, colonies, walls, borders and carceral archipelagos but also a geographical ontology of difference. Today that difference is constantly re-enacted in discourses on climate change, environmental catastrophe, terrorism and poverty in the Global South, as if the 'place of civilisation' bore no connection to the space in which the crisis occurs, nor any responsibility (Giuliani 2017b). Elizabeth Povinelli (2016) has interpreted this ontology as a geontology constructed by geontopower that projecting non-human and non-human-enough life onto the realm of Nonlife rearticulates within neoliberalism the opposition between 'we' and 'they' that is fundamental to politics and biopolitics. Within this frame, aliens are beings labelled as Nonlife that humans can either decide to exterminate or re-think as bound to them by a mutual connection (pp. 49–50). Kathryn Yusoff (2018) has reread the realm of Nonlife in connection with black and brown lives sacrificed to what I would call the 'ontologies and logics' of the Anthropocene/Capitalocene: the black and brown bodies of today's 'migrant aliens' are telling the 'we' a different story about the Anthropocene and the apocalypse (see also Section 1 of this book) – a story that is often acknowledged or rejected, denied and violently silenced.[1]

Filmic narratives of alien invasions construct 'outer space' as an Otherworld, that is, an alien space having no apparent connection or communication with the human world. Uncoincidentally, *communication* with aliens has always been a key theme in alien films. Hindered by *bordering*, *Outsiding* and *Other(world)ing* practices, communication represents indeed the possibility or impossibility of bridging the distance – that is, crossing bodily, cultural, political and geographical borders (Ahmed 2000) and creating proximity based on (at least ideally) mutual recognition. As we have seen in the first section of this book, rabid monsters do not communicate but assemble in flocks, driven by the same 'instinct' due to decreased brain function. Unlike them, *aliens* are imagined as having a culture of their own and a gaze on things that is even more rational and effective than humans'. But the purported absolute strangeness of their symbolic system and semiotics means that by 'staying', as Simmel reminds us, *aliens* could potentially undermine the 'we's' bodily structure, way of thinking, behaviour, lifestyle and social organisation or even introduce new forms of life. While societal fears of aliens 'staying' and eroding or destroying the foundation of the 'we' are widespread, as Giovanni Sartori (2000, 2015), Oriana Fallaci (2006) and Slavoj Žižek (2016) remind us with their arguments against letting refugees into Europe, the possibility that they might foster new forms of life is generally dismissed as unrealistic and confined to post-apocalyptic film narratives.

Fears of invasion, in the sense of Hage's 'état de siège' (Hage 2016; see also Lochak 2007), are widespread in Europe, where there is a general perception that the people, history and lands that were artificially separated and kept at a distance are now overflowing its space. They cannot be seen as simply 'coming', having been constructed through *Othering* and *distancing* processes. Rather, they are seen as obliterating the artificial distance enforced by human politics, economics and geographies while they are in fact *re*-connecting, bringing together what was previously cut, severed, and violently Othered by what we may call the Western *cartographic reason* (Farinelli 2009; Harley and Woodward 1987; Sparke 1998; Olsson 2007; Huggan 2008; Gregory 1994). As Reece Jones summarises in *Violent borders*, 'maps, hedges and fences [continue] enclosing the commons and bounding the sea' (2016, p. 183). State borders are everywhere, not only within and around Europe, but in accordance with my politics of location, my aim is to explore the monstrification of border crossings that sustains an imaginary of fear in Europe and the broader West. It is from my location that I try to understand why migrants' troubled geographies (the troubling implications of their mobility towards Europe) make their mobility appear in the mainstream narrative as a deadly threat subverting the order of things.

This section opens with a brief history of *bordering*, *Outsiding* and *Other(world)ing dispositifs* in Europe and the West. I will examine their connections to the biopolitics of monstrification singling out the bodies in motion that have been fleeing in search of freedom, individually or in groups, after experiencing varied and intertwined systems of subjection during Modernity (Chapters 2 and 3). Building on my interpretation of current fears of invasion as the vengeful return of repressed memories of violence, I will argue that narratives about racialised, autonomous and unruly bodies in motion shape European and generally Western self-image and concerns about historical identity, role in the global balance of power and relations with the environment. I will trace the history of alien encounters and invasions in science fiction, exemplifying, respectively, the (white) narrative of progress (the white man's journey to the Moon) and the backlash of progress (the monstrified alien invasion). I will also stress the epistemic relation between images of the invasion from Outer Space (and the unruly mobility of the modern subaltern) and current 'fears of invasion' fuelled by mass migration to Europe and by the crisis of the Anthropocene's ontologies and logics.

Fears of invasion are not new, although their origins are often traced to anxieties aroused by globalisation, the war on terror and its legacies (Brown 2010). Fears of social and political unrest or colonial counter-invasion were socially constructed and politically mobilised during Modernity to limit the right to spatial and social mobility and place restrictions on free movement across all kinds of borders and boundaries – be they defined by race, class, gender and sexuality, free or enslaved status, domesticity, family and the household, private and public space, the State's territory, the metropole and the colony. The right to mobility is intimately related to the spatial organisation of surplus-value extraction and depends on social status and privileges across not only racial, gender, class and religion but also geological lines, because, as Kathryn Yusoff argues, 'geology is a mode of accumulation, on one hand, and of dispossession, on the other, depending on which side of the geologic colour line you end up' (2018, p. 177).

Drawing on critical border studies and environmental humanities, a postcolonial analysis of contemporary biopolitics of *bordering*, *Outsiding* and *Other(world)ing* will lead to developing a notion of the border as a semiotic *dispositif* that, at the same time, is sustained by and reproduces ideas of monstrosity (Chapter 4). Through a historical contextualisation of socio-political dynamics and narratives, I will highlight continuities and fault lines between the past and the present leading to current rearticulations of the biopolitics of *Othering* and *distancing* (Asad 1973; Clifford 1986; McGrane 1989; Fabian 1992; Trouillot 1991; Ahmed 2000) in discourses and practices of securitisation defining the 'frame of war' (Butler 2009) that since 9/11 have shaped political and cultural debates on people's mobility and the terrorist threat in Europe and the West. Chapters 5, 6 and 7 analyse film narratives to explore mainstream and alternative imaginaries of the *strange encounter* (Ahmed 2000) with the *distanced* Other through the allegory of the alien invasion. Visual news media and films instantiate the legacy of cultural (national) and colonial archives of figures of race in motion. They can help deconstruct the symbolic material sustaining today's camps, colonies and borders and provide alternative conceptions of space, encounter and communication that may allow re-conceptualising *geography* along critiques of the Anthropocene and Capitalocene by reducing the distance.

Several of these alternative conceptions explore the possibility of a *partage* (Iveković 2019) with what is regarded as an *absolute* Other, pointing to a future other than *ultimate war* and total annihilation. They are the British films *Monsters* (2010), directed by Gareth Edwards, and Jonathan Glazer's *Under the Skin* (2013), which I will contrast with Neill Blomkamp's *District 9* (2009), and the U.S. productions *War of the Worlds* (2005), by Steven Spielberg, *The Mist* (2007), directed by Frank Darabont, Denis Villeneuve's *Arrival* (2016) and Alex Garland's *Annihilation* (2018). While communication and proximity are not even an issue in *War of the Worlds* and *The Mist*, in *Under the Skin* the she-alien has learnt English to better hunt humans. In *District 9*, alien language is treated as one of the many spoken in South Africa, while in *Arrival*, it is translated and used by humans to reverse their perception of time. In *Annihilation*, alien 'bio-semiotics' reassemble bodily identities, making humans and non-human entities communicate *materially*.

As in the rest of this book, I analyse films and all visual texts as assemblages of narratives that I use to build and support my argument. My reading thus offers a rather selective, if not reductive, picture of their textual complexity, one that serves the purpose of deconstructing the we/they *dispositif* that legitimises dichotomic readings of human and non-human phenomena and their relation with *geography*. Films are neither case studies nor texts that encapsulate a hypothetical truth: on the contrary, they are theoretical clusters that I interpret as philosophical clues and use to develop my argument.[2]

More specifically, I take them as exemplifications of the biopolitical regimes established to deal with the *strange encounter*: segregation and elimination, co-existence and empathy, and the unsettling communication between humans and non-humans (see Chapters 5, 6 and 7, respectively). Films are especially helpful in clarifying two aspects of the 'we's' imaginary of invasion. First, they provide different answers to a key 'question': What do they want from us? In line with the genre's conventions, in the films I examine here aliens possess a much more advanced knowledge than humans, but they are visiting Earth because they need something. They can be predators (*War of the Worlds* and *Under the Skin*) or monsters called into an unknown land (*The Mist*); extraterrestrial creatures who, like migrants stranded at the border, get trapped on Earth (*Monsters*) where they are confined to camps and slums (*District 9*), or visitors from the future (*Arrival*) and makers of a new world (*Annihilation*). A deadly invasion is the common theme in *War of the World*, *The Mist* and *Under the Skin*, with aliens respectively attacking, striking back in self-defence, or exploiting humans. That is not the case with *Monsters*, *District 9*, *Arrival* and *Annihilation*, where the alien invasion, whether by accident or on purpose, is meant to change the way the 'we' conceives of the world to the point of feeling *annihilated* – the invasion does not bring death but a deep transformation of *geography*, *Time* and *biology*. There is a striking parallel between the key question in these films and the public debate on the so-called migrant and refugee 'crisis'. After 9/11, the 'we' stopped wondering if migrants and refugees were coming in peace, and the question became: Why do they hate us? Eventually, it led to the war on terror against Al Qaeda and its allies and new practices of 'homeland security'. This 'question' represents the link between present fears of invasion and the colonial and cultural (national) archives of panic – caused by the perceived state of siege, fears of losing national identity and conquered lands, contamination and miscegenation that in the metropole, the colonies and the postcolonies shaped the *strange encounter* with the *distanced* Other, situating it within 'violent geographies' (Gregory and Pred 2006; Hage 2016; Latour 2018, pp. 24; Giuliani 2016a).

In Europe and the West, as Wendy Brown acutely argues, 'the figure of alien danger is thus literally overdetermined today, comprising economic, political, security, and cultural effects of globalisation. These disparate elements are fused into one, producing "the alien" as a many-headed dragon' (2010, p. 117). There is, however, another dimension to this figure, and that is *bio-mutation*. The potential loss of humans' bio-racial identity at the hand of the alien (as if white race and normativity defined humanness) triggers two very different reactions. On the one hand, there are fears of a reverse postcolonial genocide – that is, a backlash of the impact of modern colonisation and Western viruses and bacteria on the indigenous people. Likewise, there are fears that an alien invasion will result in the extinction of the former coloniser. On the other hand, there is enthusiasm for a new, beautifully monstrous dimension that is contrapuntal to the Anthropocene, which Donna Haraway has called the Chthulucene:

> Chthulucene is a simple word. It is a compound of two Greek roots (*khthôn* and *kainos*) that together name a kind of timeplace for learning to stay with the trouble of living and dying in response-ability on a damaged earth. *Kainos* means now, a time of beginnings, a time for ongoing, for freshness. Nothing in *kainos* must mean conventional pasts, presents, or futures. [...] Chthonic ones are monsters in the best sense; they demonstrate and perform the material meaningfulness of earth processes and critters. They also demonstrate and perform consequences. Chthonic ones are not safe [...] No wonder the world's great monotheisms in both religious and secular guises have tried again and again to exterminate the chthonic ones. The scandals of times called the Anthropocene and the Capitalocene are the latest and most dangerous of these exterminating forces. Living-with and dying-with each other potently in the Chthulucene can be a fierce reply to the dictates of both Anthropos and Capital.
>
> (Haraway 2016, p. 2)

The second aspect concerns the biopolitics of defence and its allegorical significance in these films as in those about *fast-running* and *devouring* posthumans. The biopolitics of defence represents either an allegory of existing military and governmental *dispositifs* such as camps, colonies, border, walls, and carceral archipelagos (Agamben 1998; Brown 2010; Mezzadra and Neilson 2013; Stoler 2016, p. 77) established to control migrants flows and the terrorist threat, or that of an alternative *modus vivendi* based on co-existence and translation/communication. Postcolonial biopower operating in Europe and the Mediterranean today through the civil, military, and security industries (Lemberg-Pedersen 2018; Cranston et al. 2018; Andersson 2018) is structured by the apparatuses described by Agamben, Brown, Stoler, Mezzadra and Neilson. Camps, colonies, walls and borders are clearly central themes in most of these films. The wall built to 'contain immigration from South America' in *Monsters* delimits the red zone where aliens landed and proliferate. In *The Mist*, a supermarket becomes the colony that protects humans from aliens, and the military base in *Annihilation* is a bulwark against the threat of the ever-expanding alien Shimmer. *District 9*'s segregated camps (slums) constrain alien mobility, while *Arrival*'s alien spaceships epitomise the border between human and non-human. In *Under the Skin*, a house has been turned into an alien camp where the she-alien transforms humans into nourishment; in *War of the Worlds*, the disappearance of such camps and borders is the reason for waging a war to the death against the invader.

Alternative narratives are offered in *Monsters*, *Arrival* and *Annihilation* where, respectively, a silent proximity of humans and aliens allows for the possibility that Life and Nonlife may co-exist. In contrast to the geological dichotomy sustaining the Anthropocene, language-based communication lets humans modify their perception of time, and DNA recombination generates the beautifully monstrous Chthulucene where, as argued by Donna Haraway (2016), time, space, Life and Nonlife also recombine.

2 A history of *Outsiding*

As Foucault (1961) highlights in *Histoire de la folie*, Western societies have always ostracised that which was perceived as threatening (Nuzzo 2013, p. 56). Foucault's periodisation (1999) of Western *Outsiding/Othering* of political monsters (political criminal offenders in ancient times; deformed, repulsive misfits, or 'natural monsters' in the Middle Age; and common criminals, or 'moral monsters', since the eighteenth century) reveals how the figure was used as a *dispositif* to demonise both the internal abject and the *New World*'s indigenous people, along with their 'non-normative' knowledge (Dussel 1995; Povinelli 2016). In line with Kristeva (1980), I see abjection as something that 'disturbs identity, system, order', but which forms an indispensable part of the self. I see the 'self' she refers to as the body politic – be it that of the nation, Europe or the West – and the 'abject' as its constantly reproduced margins. Witch-ed, queer, deformed, disabled or poor people as well as unruly peasants, heretics, Jews and factory workers were long represented as criminal monsters (Sontag 1978; Olson 2013; Shildrick 2002), and as such they were disciplined or suppressed (Foucault 1976; Federici 2004), providing the negative standard against which social and moral norms were defined.

Constructions of monstrosity served specific purposes, although the complex symbolic texture of those representations and their different social uses signal varied relations to the liminal, the challenging, the uncanny, the marvellous, the frightening and the horrific. In general, monstrosity and the upholding of whiteness, masculinity, heterosexuality and land ownership as normative standards set the stage for a segmentation of European societies, colonies, and nations that legitimised gender and race-based exploitation or extermination of the internal abject and colonial subject across seas, states and imperial borders, while defining and reproducing privilege and the ontologies and logics of Anthropocene (Povinelli 2016; Yusoff 2018). Like imaginative translations of past 'figures of race', these representations constantly recombined, cross-referencing colonial and European iconographies of monstrosity, local cultures and contextual discourses to the point of being co-construed. A well-known example of this is the interrelated monstrous construction of black people and the feminine that ultimately inspired the figures of extra-terrestrial monsters such as the xenomorphic mother in Ridley Scott's 1979 film *Alien* (Creed 2000, pp. 110–135; see also 1993).

Well exemplified by the black dog-headed body (the Muslim *cynocephalus*), the figure of the *alien-ed* monster circulating since ancient Rome, early Christianity and the Crusades connected deformed, animalised monstrosity to such transgressors of the 'order of discourse' as criminals, infidels and heretics in both the Holy Land and the domestic space of the witch (Foucault 1999; Lucchese 2019; Friedman 1981; Federici 2004; Arjana 2015; Wolfe 2005; Mittman and Dendle 2012; Mittman and Hansel 2018a, 2018b). Monstrous animals, infidels and heretics as well as sorcerers and witches were thought to be coming 'from elsewhere' – not yet outer space, but rather the realms of Nature, Chaos, Darkness and Evil allegorising all that was regarded as foreign to the *polis*, Christianity or God's word and, from the eighteenth century onwards, Western civilisation.

During the European Renaissance, travel diaries and bestiaries portrayed a New World riddled with monsters, according to a necessarily binary vision separating light and the beauty of creation from repulsive, horrific otherworldliness. Africa, the Caribbean, Amazonia and the Pacific were inhabited by Calibans who, like Shakespeare's misshapen half-man half-fish (Shakespeare 1610–1611) could be educated only to some degree, because their ontological diversity, signified as monstrosity and brutality, was regarded as an immutable fact (Schmigall 1981, pp. 176–177).

Seventeenth- and eighteenth-century scientific approaches turned every individual, group and organisation seen as socially unruly into something *naturally* doomed to imperfection and chaos. The state of nature hypothesised by Hobbes and Locke in the colonies and in seventeenth-century Europe shaken by religious wars, as well as the Hobbesian Behemoth of regicide rebellion in the colonial metropole and the revolutionary *Tiers État* imagined as a changing force by French and British political philosophers (Abbé Sieyès, Jean Bodin, Jean-Jacques Rousseau, and Edmund Burke), implied a threat to the political order of monarchy and the higher social ranks. The threat came from the indigenous populations, the enslaved, and the human mobs depicted – just like pirates and fugitives in the seventeenth and eighteenth centuries (Linebaugh and Rediker 2000) – as many-headed hydras (Neocleous 2005, p. 33).[3]

In this context, even unknown or unmasterable *natural* elements, phenomena and entities were interpreted as otherworldly, and the otherworldly as threatening and monstrous. Being against reason, they signified disaster and apocalypse. Since the beginning of the age of technology and industry, ostracised outcasts that capitalism could not annihilate or make productive had been labelled as *monstrous* – as if machines, the evil capital-led, worker-swallowing Moloch described by Marx (1867), epitomised the new era of

Figure 2.1 Francisco Goya, 'Witches in the air' (fragment), oil on canvas, 1798.

absolute reason and rightfulness. All that was situated outside dominant capitalist and colonial racist and sexist bio- and necropolitics was regarded as the *monstrous* elsewhere – a construction underpinned by multiple, intertwined, centuries-old discourses and rearticulated in Western Modernity within the

Christian and colonial dichotomy between God and Evil, civilised and barbarous, productive and destructive, Life and Nonlife.

Nineteenth-century positivist discourse and deterministic theories turned race, class, gender, and sexuality into measurable entities, resulting in the criminalisation and animalisation of racialised and gendered groups, and racialisation and genderisation of the lower classes (Olson 2013). These monstrified subjects were believed to be justification for draconian laws, severe punishments, mass segregation, pauperisation and slavery. Their rebellion and unruly mobility – as we have seen in Section 1 in the case of slave mutinies and revolts – crossed social boundaries and the walls protecting the living space of the privileged, triggering *moral panic*. As such, they were generally described as *alien* invasions from the *finis mundi* leading to social and political chaos and the end of the *world as the 'we' wanted it*.

Modern practices of sovereignty, discipline, power and biopolitics targeting the *alien-ed*, *alienised* and criminalised unruly monsters were established to let progress shine, keep the pace of capitalist acceleration and allow actual or imagined colonisations to encompass the geography of the whole universe.

It was within industrial Modernity's practices of abjection and *Outsiding* that the figure of the alien living on another planet or in another time-space dimension emerged to become a popular topic in science and science fiction, although its literary origins can be traced back to antiquity and theories of cosmic pluralism. Early twentieth-century novels, short stories and films focused on remote or interplanetary expeditions and man's encounter with the extraterrestrial. The Moon was the subject of popular literary works such as Herbert George Wells' *The First Man on the Moon* (1901), Edgar Rice Burroughs' *Under the Moon of Mars* (1911), French astronomer and sci-fi novelist Nicolas Camille Flammerion's *La pluralité des mondes habités* (1862) and *Les mondes imaginaires et les mondes réels* (1864), and Italian writer Emilio Salgari's *Le meraviglie del 2000* (1907). It was a popular theme in cinema too, with pioneering French film director Georges Méliès' *Le voyage dans la lune* (1902) and its remake by Segundo de Chomón, *Excursion dans la lune* (1908), Méliès' *Voyage à travers l'impossible* (1904), Percy Stow's *When the Man in the Moon Seeks a Wife* (1908) and Enrico Novelli's *Un matrimonio interplanetario* (1910).

Fantasies and fears rendered as alien invasions first appeared timidly at the end of the nineteenth century. The genre was inaugurated by British politician and writer Edward Bulwer Lytton's *The Coming Race* (1871) and Herbert G. Wells' famous *War of the Worlds* (1898) and reached its golden age during the Cold War.[4] Beside the never-ending *Star Trek* saga (1966–1969; 1987–2005; 2017–), more focused on interstellar voyages and wars, from the 1970s and especially from the 1980s onwards the genre flourished again with U.S. films like Steven Spielberg's *Close Encounters of the Third Kind* (1977); John Carpenter's *The Thing* (1982) and *They Live* (1988), Joe Dante's *Gremlins* (1984), Stephen Herek's *Critters* (1986), John McTiernan's *Predator* (1987), Roland Emmerich's *Independence Day* (1996) and TV series such as Kenneth Johnson's *Visitors* (1984–1985). Although Ridley Scott's *Alien* (1979) is not about alien invasions, the plan to return a sample of the alien to Earth can be seen as an omen.

As early sci-fi novels show, the idea that men could get to know, visit and conquer new (and superior) worlds was mostly limited to affluent white male artists or scientists. As representatives of human progress, they were entitled to move across geographical and planetary boundaries; a potential invasion from those same worlds, on the other hand, was feared as monstrous, a deadly threat to humanity. We may speculate that the contrast between the gentle landing of the white man on Sirius and the fear of man-slaughtering Siriusians colonising Earth is related to the selective memory of Western colonialism. Nonetheless, what interests me here is the parallel between representations of 'hordes' of aliens trespassing the geological boundary between human and non-human and representations of 'hordes' of 'wretched of the Earth' (Fanon 1961) rebelling against political and social borders.

In general, as Gregoire Chamayou has outlined in his work on the hunting of rebels on the run (2010), mobility across the borders of states and nations has always been defined by historical privileges, and hence the subaltern's mobility was always seen as threatening (Simmel 1908; Sassen 1999; Mezzadra 2001). This is clear when looking at how mobility was regulated across time, along with the foundational elements of Western society that have become the paradigm of modern colonial capitalism. Among them are the industrial factory (Marx 1867), the school, the mental asylum, the penal colony and the prison (Foucault 1975), family and the household (on women, slaves and servants, see Federici, 2004 and

Figure 2.2 Still from *Monsters* (2010), directed by Gareth Edwards. Samantha (Whitney Able) and Andrew (Scoot McNairy) look at the US–Mexico border wall from the top of a Mayan pyramid.

Stoler 2002), the plantation (Allen 1994; Hartman 1997; Sharpe 2009), and the slave ship (Rediker 2007). Through these disciplinary *dispositifs*, bodies have been mastered, contained and confined or forcefully mobilised since early capitalist and colonial Modernity across spaces of Christianity/civilisation and spaces of barbarity. But from the fringes of disciplinary society and the Out there – the *Hic sunt leones* or the place of Evil, where nature is (still) wild and rebel – transgressors of all sorts of borders, camps and colonies for racial segregation (e.g. for the Roma and the Jew in Europe, and for the colonised in overseas dominions), sex workers, and waged or enslaved workers emerged to become dangerous monsters (Linebaugh 1991; Moulier Boutang 1998; Mezzadra 2001). As Wendy Brown has shown for the present time, building on an idea of borders and walls as biopolitical *dispositifs* that make the border-crossing individual monstrous, 'walls against immigration construe the trespassing as an "invasion"' (2010, p. 129).

As confirmed by artists and art and history scholars investigating colonial and pro-slavery representations of the monstrous body (see also Section 1), the bordering practices and monstrous intimacies of slavery constructed the enslaved monster as a 'criminal' while at the same time accomplishing their own legitimation, inasmuch as the enslaved would inevitably engage in rape and murder if not confined. I am thinking in particular of Christina Sharpe (2009), Kara Walker (2001, 2014), Janell Hobson (2005), Glenn Ligon's *Runaways* series (1993), Magdalène Dumas' *The Image as Burden* (1995) and Renée Valerie Cox's *Venus Hottentot 2000* (1995). Practices of immobilisation and constraint were thus used as *dispositifs* to hide the violence of slavery and project meanings of violence and moral chaos onto the enslaved/colonial subject.

Figures of emancipatory mobility have been depicted as beastly (Olson 2013) and causing *moral panic* since early modernity. Among them were the wandering vagabonds escaping social control and the discipline of work and exploitation in the colonies (Linebaugh and Rediker 2000), plantation fugitives and *quilombolas* (Allen 1994), women challenging patriarchal constraints (Irigaray 1977; Braidotti 1997; Federici 2004; Santos 2016), paupers fleeing the poorhouse, and segregated colonial populations escaping environmental disasters. The same fate befell slave mutinies at sea, anticolonial struggles, breaches of apartheid's social and physical borders, and poor people and minorities' emancipatory migration within and across former colonial peripheries and metropoles.

All these figures, the dark face of the 'bright' and 'progressive' era of the Anthropocene, have contributed to constructing the meaning of monstrosity – along with that of security, morality and normativity in Western countries and their colonial peripheries – and the privileged wellbeing of the normative Anthropocenic Anthropos (Povinelli 2016; Haraway 2016; Yusoff 2018). The same moral distance imposed upon the environment, by virtue of which beyond the borders of civilisation lies the place *of* disorder and disaster, extends to monstrified border transgressors, seen not just as threatening bodies but as the medium through which social and environmental disasters typical of the elsewhere can reach the space of the 'we'.

As with the alien monster in industrial Modernity, they are seen as opening the door for Chaos to enter the increasingly precarious order of the Anthropocene, signifying the deadly effects of neoliberal acceleration and proximity to the Out there that capitalist expansive extractivism is producing. Just as Wells' invasion was a reminder of the backlash of modern colonisation, so too alien invasions in film seem to evoke the unintended consequences of the Anthropocene. This is clear in *The Mist*, where aliens land on Earth through a 'door' opened by the army between worldly and Otherworldly dimensions, and in *Monsters*, where a NASA space probe lands back on Earth bringing with it alien spores. The belief that Western science and technology may have mitigated proximity to chaos in the name of progress, re-production, national wars and power gradually yields to the uncertainties epitomised by the racialised and uncivilised. Today:

> at the same time that racialised discourses justifying colonialism, natural hierarchies, and global inequalities have lost their easy hegemony, global movements of people and capital have eroded the separate spheres inhabited by the populations these stratifications produce. Today, rich and poor, colonizer and native, First and Third World live virtually and actually in ever greater proximity.
>
> (Brown 2010, p. 122)

In order to reproduce the dichotomous view of the world divided between the civilised and the barbarians (Amoore and De Goede 2008, pp. 5–6; see Section 1, Chapter 6 of this book), the Out there needs to be contained and secured (also as a way of safeguarding extractivism and exploitation), and the 'illegitimate' and 'uncivilised' *risky bodies* (Aradau 2004) flowing into Europe need to continue being *distanced*, 'dispensable', and exploitable both within and outside the receiving countries. In brief, the same distinction between citizen and subject, a legacy of colonialism that according to Mahamood Mamdani (1996) keeps Africans subordinated to undemocratic forms of rule, needs to be kept alive under different conditions in Europe and the West. As Ann Laura Stoler (2016) maintains, these *dispositifs* and distinctions are the marks of:

> [Colonial] duress [which] I conceive it is a relationship of actualized and anticipated violence. It bears on those who are its perpetrators, produces anxieties, and expanding definitions of insecurities that are its effect, a demolition project that is eminently modern, and as Frantz Fanon conceived it, a form of power that slashes a scar across a social fabric that differentially affects us.
>
> (p. 8)

In Europe, the government of people and their mobility operates along colour and gender lines that define the space of citizenship and through a range of practices that regulate and reproduce the lives and deaths of both the normative 'we' (that is, white, affluent, debt-free and generally hetero-normative full citizens) and the *risky bodies* from whom they need to be protected.

In a geopolitical landscape that since 9/11 has connected overseas scenarios of the war on terror to border regimes, homeland security and social control, these practices also set the levels of violence that sovereign institutions (the state and its army, police and government apparatuses) should use in the 'management' of risk.

Unlike the invading alien or zombie hordes in science fiction, the Other who comes and the Other 'within' – that is, the stranger who stays, the home-grown alien – cannot be annihilated, or at least not in the same way. Hence the border restrictions, racial profiling, containment, walling, deportation, detention and other deadly practices that prevent the 'killables' from contaminating the 'legitimate' with their diminished humanness in much the same way that an alien or rabid monster would do.

3 *Othering, bordering* and *Other(world)ing*

Differential entitlement to mobility has undergone significant changes across Modernity. Today, mobility as a person's universal right (1948) 'to leave any country, including their own, and to return to their country' must contend with official, procedural and symbolic restrictions as well as with discursive and material constructions of the subjects of mobility. The differences between past and present ways of controlling or regulating mobility are many and include the emergence of new actors: the state first and then the nation as main agents of control have been joined by governmental and non-governmental agencies, private companies, and supranational or international institutions. Other differences revolve around the legal status of the persons in mobility, who are no longer colonial subjects but officially free post-colonial citizens; their trajectories, heavily constrained by forced im/mobility in the past and nowadays also affected by the practice of bouncing migrants from place to place described by Tazzioli (2017); and the multiplication of borders and their apparatuses in a supposedly border-free globalised world, as outlined by Mezzadra and Neilson (2013).

Today, as during Modernity, the reasons that drive people to move are connected to the Anthropocene. There is an important difference, though. Geontopower, as Elizabeth Povinelli (2016) calls it, and the (ge)ontological dichotomy of Life vs. Nonlife – including in the latter inanimate rocks, metals and inorganic natural resources – have historically been the foundation of colonialism and global capitalism. Our times are instead marked by what Povinelli calls a neoliberal 'acceleration' of the impact of the Anthropocene and Bruno Latour (2018) the 'New Climate Regime'.

Because of its global dimension, neoliberal acceleration takes geontological power (geontology) to the extreme. The dynamics of neoliberal acceleration have led to extreme extractivism – with its appropriation and devastation of Nonlife – as well as to the escalation of violent governance practices, warfare, biopolitics and necropolitics (Foucault 1975). Even the war on terror, aiming to eventually control and extract value from natural resources in the Middle East, should be regarded as a consequence of this acceleration. Climate change, and the unbearable increase in global emissions, is the final outcome of this geontology. The 'objective causes' behind the displacement of peoples and the choice to migrate (Mezzadra 2001; 2004, p. 270) can be traced back to the acceleration of neoliberal capitalism, which also exacerbates primitive accumulation, conflicts, wars, scarcity of freedom, discrimination, poverty, famine, epidemics, pollution and environmental disasters (Armiero and Tucker 2017).

The same biopolitical *dispositifs* governing other mobilities from the so-called Global South – borders, walls, camps, colonies and carceral archipelagos – prevent migrations driven by climate change and environmental catastrophes from physically and symbolically contaminating the colonial metropole (Bettini 2013, 2017). The imaginary sustaining such *dispositifs* is that of a homogeneous, colonial, European and Western Self, while the West's elsewhere – wherever that may be – is uniformly constructed (for instance, by institutions dealing with climate change) as the place *for* disaster. As I have anticipated earlier in this book, the distinction between the Self/inside as a place *of* disaster and the Other/Outside as a place *for* disaster recalls the colonial distinction between civilisation and barbarity, or between those able to 'master' nature and those who are overwhelmed by it (Giuliani 2017a, 2017b). To better understand this distinction, I will associate these constructions with the concepts of catastrophe and disaster, respectively. As I have argued in the introduction to this book (see also 2017b), catastrophe is understood as a crisis, an event or a moment of transition occurring in the West. Catastrophe 'ruptures the continuum of time and liberates both positive (Benjamin's evaluation of the time of revolution) and negative imaginaries oriented towards futurity' (Baishya 2011, p. 2). A disaster instead is seen as an uncontrollable event occurring elsewhere in the world, outside Western space, where barbarity is systemic. This dichotomy is part of what Wendy Brown (2017) has described as the 'indispensable yet intensely colonial trope [ontogeny recapitulates phylogeny] for framing civilizational infancy and maturity, difference and not only development, [which] has for millennia organised our understanding of the human and its possibilities' (p. 28).

Hegemonic narratives of catastrophe work in the same way as hegemonic narratives about endemic terrorism, epidemics, wars and poverty. They separate events and phenomena from their local and global contexts and interrelated causes and effects, treating disasters as if they were isolated events with no mutual relation (as if the effects of climate change could be seen in isolation) and the elsewhere as *the* place *for* disaster, revealing a discursive operation that seeks to de-politicise the causes of climate-driven migration:

> [T]he fact that an issue is depicted as a catastrophe can in certain cases even facilitate its reinsertion into the frame of normality. The fear mobilized by apocalyptic narratives on climate-induced migration can act as the traumatic element favoring a (re)normalization of the issue, passing from denial to trivialization, from impossible to real element of business-as-usual, to be managed by governance instruments. Narratives on climate-induced migration that foresee hordes of wild, destitute barbarians menacing the wealthy, do not necessarily imply that something 'radical' will be done. These alarmist narratives may eventually provide legitimacy and consent to political options that do not avoid disruptive ecological changes but rather frame these outcomes as unavoidable 'part of the game'. As a source of legitimacy for policies that deal with climate change without affecting the social, ecological and economic relations that set the scene for anthropogenic climate change.
>
> (Bettini 2013, p. 69)

As a result of this discursive strategy, what attracts international public attention is the 'rescuing' of 'deserving' climate-change refugees rather than the need to systematically question the workings of capitalism, *reveal* its effects on the planet, and find radical solutions. Rather than a responsible political approach to the causes, a new kind of humanitarian (civilising) mission is privileged, aimed at selecting the ones deemed worthy of rescuing while softening anxieties tied to fears of invasion and climate apocalypse.

Fears of invasion combine with pity for the dead and empathy for the refugees in the current iconography of migrant landings, and in the name of a new moral economy centred on the humanitarian reason (Fassin 2011), the plea for aid to the refugees is coupled with the demand for tougher border-enforcement measures against 'economic migrants'. The interrelationship between fear, pity and empathy, critically described by Hartman (1997), Saucier and Woods (2014) and Danewid (2017), turns the 'poor victim' fleeing danger into the recipient of the 'we's' benevolence and superiority (Cole 2006). This cleanses 'us' of our guilt for turning a blind eye to the violence of borders (Aas and Gundhus 2015; Jones 2016, pp. 57–68; Giubilaro 2017, 2018) and its underpinning ontology, or geontology, which reduces some human beings to objects of disaster. In the (post)colonial discourse of the humanitarian (civilising) mission, this implies shifting the blame for that violence onto wars waged by fanatic barbarians, or sudden catastrophes against which man is powerless or, finally, onto migrant smuggling. In the eyes of the 'we' who has the right to look (Mirzoeff 2011) at the threat and define it, climate-driven environmental disasters are either caused by humans' misdeeds (hence by the lazy barbarian, the fanatic or the cruel capitalist) or they are fortuitous events, rather than the result of complicated interactions between humans and the environment (Barad 1995; Tuana 2007). Similarly, shipwrecks off the coasts of Lampedusa and Malta are ascribed to bad weather and the dangers of sea crossings rather than to the effects of bordering (Giuliani 2017a).[5]

The blame-avoiding narrative surrounding humanitarian acts of compassion renders the very object of compassion 'voiceless' and at the mercy of national and European guidelines for verifying the authenticity of their stories. This situation is reversed in the film *Arrival*, where the alien landing is motivated by the desire to avert a catastrophe for which human help is required. The monstrified alien here can speak and write and tells the story of a threat looming in the future that may be averted through collaboration and communication between humans and non-humans, provided that humans accept the 'gift' of alien language and non-linear perception of time. If we read this film contrapuntally through Michel de Certeau's (1975) critique of colonial historiography then the purpose of the alien gift is to transform the reassuringly linear representation of *Other(worldly)* temporalities in Western (colonial) history-writing. The linear perception of time embedded in Western ideas of progress is seen as doomed. The apocalypse feared by the aliens can be interpreted as an allegory of climate change effects; similarly, the *alien* risky body coming from elsewhere that, akin to a Sybil, predicts a common tragic fate and provides a way to avoid it through collaboration can be seen not only as the allegory of an embodied critique of modern and colonial hierarchies of space, but also as a plea for collaboration in undoing the workings of the Anthropocene. The monstrified alien, in a way that is not dissimilar to Benjamin's (1940) angel of history, represents the future that looks back for a solution and in so doing *reveals* and erases the distance between human and non-human, reality and otherworldliness, Self and Other, past and future. Essentially, it nullifies the workings of the semiotic power of the border or, better, reveals its illusion of separateness and, as Graham (2002, p. 12) indicates in the opening quote to this section, its own instability.

Sci-fi cinema has widely explored the inherent instability of the border. From Wanger's *Invasion of the Body Snatchers* (1956), based on Jack Finney's novel (1955), to Romero's *Land of the Dead* (2006) and Blomkamp's *District 9* (2009), the separation between humans and non- or post-humans is always described as ephemeral. While Wanger's aliens take the appearance of the humans they replace, Romero's zombies gradually acquire feelings, memory, skills and a sense of community. In *District 9*, aliens replace the poorest locals and immigrants in the 'black' slums of Johannesburg. The border *bridges diversity*, or what appears as such. In *Annihilation* the border's instability is also *bio*logical – which is why I chose to interpret it as both a *bio*political bridge and a *bio*political fault line – given that aliens have the power to recombine human and non-human DNA and language.

In Chapter 6, I will discuss post-human theories in connection with this film to better describe the border as separating/connecting *zoe* (Giuliani 2015, 2020).

As I have argued elsewhere (2016b), Graham's concept of 'ontological hygiene of the human subject' (2002, p. 12) works at both a bodily and social level. The (human) body and the body politic have long been considered to be in danger and in need of 'eugenic' preservation from the hordes of aliens that, like a mortal disease, have spread to the heart of Europe. The apocalypse was never too far away – and in recent times hordes of barbarians have attacked the West and its symbols: New York's Twin Towers, the London underground, Madrid's train station, Paris, Berlin and Brussels. What needs to be protected against miscegenation, contamination, degradation and ultimately annihilation today is that same society identified in popular culture as being under siege by aliens, zombies, and the environmental apocalypse. By laying claim to the same rights enjoyed by the 'we', the barbarian monster poses a threat to what was meant to be exclusive to the white normative subject: the right to certain liberties and self-comprehension and the privilege of residing in the secure spaces of the Anthropocene and pursuing a better life across borders and boundaries. At the same time, the monster instantiates the ontological difference between here and elsewhere, 'we' and 'they', that sustains and legitimises the 'we's' privilege.

A variable economy of monstrifying signs sustains the border system and its micro practices of containment, biopolitical control and racial hierarchisation, which are aimed not so much at filtering migrants but at providing the symbolic and material terrain for the extraction of surplus value from their constrained and forced mobility. Surplus-value extraction is sustained by the variable equation between the perceived degree of morality of the subject in motion and their right to mobility: the ontologised condition of the migrant comprises today notions of inferiority and animalisation and notions of inherent criminality tied to the alleged fanaticism that is associated with non-Western backgrounds. Drawn from colonial and cultural (national) archives, such notions drive the belief that migrants' mobility and 'quest' for protection or emancipation are not legitimate. By notions of inferiority, I mean the de-personalisation, dehumanisation and de-individualisation of migrants. By animalisation, I mean the process of associating figures of race with *moral panic* and an unmasterable challenge to stability, against which the colonial/metropolitan order is constantly reconstituted. By illegitimate or abusive mobility, I refer to racialised Others daring to cross the borders drawn between Europe and the postcolonial subaltern.

Sara Ahmed (2000) argues that objectivation and stranger fetishism lead to:

> 'cutting off' of figures from the social and material relations which overdetermine their existence, and the consequent perception that such figures have a 'life of their own'. Stranger fetishism is a fetishism of

> figures: it invests the figure of the stranger with a life of its own insofar as it cuts 'the stranger' off from the histories of its determination.
>
> (p. 5)

Building on Wendy Brown's (1995) argument about the fetishisation of the wound in subaltern politics, Ahmed maintains that such fetishisation 'turns the wound into something that simply "is" rather than something that has happened in time and space' (2004, p. 32). In fact, not only does the border – a social construct that is meant to separate – cut off the stranger from their constituting story, isolating them from a *geography* and a *history* of complicities that Europe and the West are also involved with, but it is also central to the semiotic construction of the entities it is supposed to differentiate. The border is both a physical and a semiotic device that constructs the Other in motion as alien, thus legitimising their geographical distancing. At the core of this *alien-ing* process is the semiotic power of discursive practices to produce the border (De Genova 2013) between the 'we' and the 'space invaders' (Puwar 2004) through reiteration (Butler 1993). The border's reiterated operations both produce and are produced by an iconography of monstrosity based on 'figures of race' that constitute at once the semiotic landscape in which the border is grounded and the symbolic material that is constantly reproduced – and thus made into 'truth' (Tagg 1988) – by the border as a technology of biopower.

Both here and in my analysis of returning memories of violence (see Section 1), the global reorganisation of the security discourse (Salerno 2016, 2017) – now facing a new phase in the evolution of the war on terror and border regimes – plays a key role in the relation between the border, the wall and the fictitious dichotomy of 'we' and 'they'. Additionally, local issues affecting political, social and environmental dynamics between Europe, Africa and the Middle East have been found to have an impact on the restructuring of borders. Among these were the consequences of the economic downturn in Europe and the West, the Arab Spring, the migrant 'crisis' in the Mediterranean – triggered by the increasingly repressive measures adopted by countries in the Horn of Africa (2007–) – internal conflicts in Tunisia (2011) and the Syrian exodus (2015–2016). Also relevant in this context is the long-standing legacy of colonial and postcolonial cultural and political encounters and exchanges as well as *ideoscapes*, *technoscapes* and *ethnoscapes* (Appadurai 1996) – global flows opposing all essentialised fixity. Within this frame, marked by identity politics, multiculturalism, culture clashes, and a *war of the worlds*, I will explore the semiotic productivity of the border, that is, its capacity to (re)produce the identity-based dichotomy of we/they incorporating the structural opposition between civilised and barbarian, victim and criminal, good and bad, object and subject of disaster, Anthropocene and crisis of the Anthropocene. Different groups of people are assigned to opposite categories depending on how their body is read; at the same time, structural oppositions reinforce the gendered, sexualised and racialised connotations of the dichotomy between the citizen entitled to free mobility and the illegal(ised) migrant, resulting in the intersectional subalternisation of the latter and legitimisation of their restricted mobility rights. Both the segregation of monsters in *District 9* and the military's efforts to prevent alien expansion at the walled border in *Monsters* describe this process.

My discussion of the semiotic power of the border in Chapter 4 will necessarily connect *Outsiding*, *Othering* and *Other(world)ing* practices as part of the vast imaginary of invasion that bridges visual representations of human mobility in news media and sci-fi productions on *alien* attacks (see Chapters 5, 6 and 7). My aim will be to explore that imaginary contrapuntally, through the symbolic material that links the memory of past colonial invasions and terrorism, postcolonial mobility and the war on terror, and fears of a reverse colonisation from Other-worlds.

4 The semiotic power of the border

As Sara Ahmed (2000) argues in *Strange Encounters*:

> The alien is hence only a category within a given community of citizens or subjects: as the outsider inside, the alien takes on a spatial function, establishing relations of proximity and distance within the home(land). Aliens allow the demarcation of spaces of belonging: by coming too close to home, they establish the very necessity of policing the borders of knowable and inhabitable terrains. The techniques for differentiating between citizens and aliens, as well as between humans and aliens, allows the familiar to be established as the familial.
>
> (p. 3)

Alien films may be seen as allegories of the encounter with the migrant monster or monstrified migrant, who are either fought (*War of the Worlds*), channelled, mobilised and quarantined (*Monsters*) or detained and differentially included in segregated camps (*District 9*). In *The Mist*, humans seek protection from aliens in a supermarket that doubles as a colony, promoting the image of the colony and the nation as fortresses that is at the root of the idea of the West as one large gated community. And yet the *raison d'être* of any such fortress, as we have seen, is to *label* the Other as such, not just containing but *mobilising* their bodies – physically and semiotically.

In fact, borders are polysemic (bio)technologies, because the constraints they place on mobility depend on the tension between the process of differential inclusion serving labour exploitation and the hegemonic reading of racialised, gendered and sexualised bodies. According to Étienne Balibar, borders are polysemic because they do not have the same meaning for everyone. They 'never exist in the same way for individuals belonging to different social groups' (Balibar 2002, p. 79), and indeed this differential meaning is essential to their functioning. Borders are designed to expose different people (i.e. from different social classes) to different experiences of law and freedom. Border law enables some to cross national frontiers, while denying others; it upholds the freedom of circulation of some, while depriving others of this same freedom. Following from this, Balibar argues that the purpose of the border is 'actively to differentiate between individuals in terms of social class' (p. 82). Borders are therefore 'instruments of differentiation'. In line with Liz Fekete's analysis (Fekete 2001, pp. 23–40), Arun Sivanandan (2001) has argued that the border system is informed by xeno-racism, a racism 'that is not just directed at those with darker skins, from the former colonial territories, but at the newer categories of the displaced, the dispossessed and the uprooted, who are beating at Western Europe's doors' (p. 2).

Balibar's idea of polysemic differential borders and Sivanandan's and Fekete's arguments on xeno-racism can be integrated with the notion that borders semiotically *produce* monsters and invaders through specific readings and rearticulations of gender, class, race, sexuality, nationality and religion. Thus, borders reinforce an ontology of diversity that essentialises the relation between the body (i.e. how it is read by power), the geographical position it occupies, its right to mobility, its capacity to follow specific trajectories when it crosses borders, and its resulting capacity for change over time. By this, I refer to the idea of a normative subject that is fully able to make choices that could imply a physical transformation as much as a radical reshaping of life trajectories.

The racialised non-normative subject who illegally crosses a border is not supposed to retain opacity or subjectivity. They are not free to define themself according to symbolic materials and experiences that are not readable by immigration and border authorities: the reason why they fled needs to be one and always the same, their body a quasi-object (void of contradictions and made into Nonlife) and their life story intelligible. Being a distanced Other is an *immutable* condition. Using Mamdani's distinction between citizen and subject, we may argue that in order to be granted access to the space of the worthy, the subject's *aspirations*, *feelings*, *ideals* and *life trajectories* need to be recognised as *transparent*, *coherent* and *fixed*. Nicola Mai's (2014) notion of 'biographical border' further elucidates the essentialist distinction between the 'we' and the *alien*:

> The process of certification of the credibility of the suffering of the migrants being both helped and controlled acts as a biographical border between deportation and recognition. The ubiquity of these bordering mechanisms is part of the proliferation of dematerialized spatial and moral borders and of new capillary surveillance mechanisms and technologies of control in neoliberal times (Graham 2011).
>
> (2014, p. 180)

Unlike the European citizen, the racialised non-normative subject is expected to present a coherent narrative, for which they are held accountable. In that sense, the migration process produces their subalternisation: stripped of whatever right to opacity they may have enjoyed in their country of origin, they are condemned to transparency, coherence, authenticity, and immutability. Sci-fi narratives of alien encounters often incorporate elements that recall biopolitical control and migrants' forced immobility: aliens are first immobilised, then vivisected; their bodies are put under the microscope and tissue is taken for genetic sampling or injected with contrast agents. Transparency is the name of the game. All about them needs to be transparent, intelligible and appropriable, fixed in time and space, lest they flee scientific inquiry and military bio-control – they are made into quasi-objects, closer to Nonlife than to Life, to paraphrase Yusoff (2018). From Spielberg's *E.T.* (1982) to the amphibious creature in Guillermo del Toro's *The Shape of Water* (2017),[6] to *Annihilation*'s (2018) aliens and humans whose bio-structure has been modified by aliens, all are scrutinised, quarantined or 'eliminated'.

There can be neither radical choices nor unexpected turns, especially when the migrant needs to meet the eligibility requirements for asylum status or trafficking-victim protection. It is by virtue of the enforced crystal-clearness and immutability of migrants' bodily and embodied life and experience that it becomes possible to officially distinguish the suffering victim (object of compassion and recognition) from the impostor or criminal (the monstrous target of moral condemnation and officially sanctioned rejection) (Ahmed 2000, p. 117; Sharma and Sharma 2003; Andrijasevic and Mai 2016; Giametta 2017; Farris 2017; Giuliani et al. 2020b). The basic question about aliens in post-apocalyptic narratives is whether they are visiting earth to seek help or to wage a war. In the former case, they are usually prepared to die – like one of the two aliens who attempt to communicate with human scientists in *Arrival* – as if to emphasise their *need* for human help. If the purpose of their visit is to wage a war, they will find that the 'we' is prepared to fight back, as in *War of the Worlds*. Either way, their motives are expected to be clear and immutable. They are expected to fit into a crystallised representation. And even when they accidentally land on Earth, not driven by ulterior motives (*Monsters*, *The Mist*, *District 9*), and their only goal is to survive until they can return to their planet (*District 9*), they – and the threat of chaos – need to be contained and spatially constrained.

As Martina Tazzioli (2017) writes in describing the biopolitics of the border, 'the Hotspot System contributes to enforce forms of containment through mobility that consist in controlling migration by obstructing, decelerating and troubling migrants' geographies – more than in fully blocking them'. An erratic trajectory of life and mobility that reconfigures the 'space' out there is imposed on migrants, as opposed to the linear trajectory of the Western hegemonic subject – rational, subjective, productive, self-crafted, self-disciplined and able to queer fixed identities (Puar 2005; Puar and Rai 2002). The assumed linear trajectory of the hegemonic subject entails a right to opacity that is grounded in a conception of the Western white, middle class, heterosexual Self as crystal-clear in its ontology. Their mobility can thus be guaranteed, made intelligible within codes of normativity that structure their conduct and trajectories. On the contrary, the 'unruly' mobility of the postcolonial subject cannot be linear and as such needs to be contained and made intelligible.

This perception reflects vestiges of the discourse on the unbearable thought (for the white Western imaginary) that a non-Western racialised subject's trajectory towards and across the postcolonial white metropole may be as self-crafted and free as that of their (post)colonial *master* (see Chapter 2). It is through the constraining of their life and space trajectories that the postcolonial space of the Mediterranean and the broader space of illegal(ised) mobility are superimposed on the space of Europe and its imagined community. The subaltern subject's self-crafted mobility is interpreted through the memory of the horrors of colonialism and feared as a retaliatory counter-invasion. As such, it triggers *moral panic*, associated with fears of penetration, vanishing identity, loss of insularity from the out there, disintegration of values and *the end of the world as the 'we' knows it and wants it to be*.

This conception of the Other from 'out there' as a *annihilating* force shares the same allegorical rendition and symbolic elements as the figure of the predatory alien in science fiction (e.g. *War of the Worlds*, *The Mist* and *Under the Skin*), although the former possesses the superior skills and abilities needed to colonise or exterminate the 'we' and is generally devoid of empathy and moved exclusively by an annihilating drive. The same feelings displayed by Modern European colonisers apparently permeate all aspects of the alien colonising effort: the *alien monster* lacks empathy for humans (*War of the Worlds* and *The Mist*) and aims at exploitation (e.g. sucking human blood) or extermination, mercilessly slaughtering and dismembering humans.

There are interesting parallels between the fictional imaginary of alien invasion and the discourse surrounding homegrown terrorism since 9/11 and especially since the recent terror attacks on Europe: in *War of the Worlds*, aliens have been living 'among us', unbeknownst to us, while an alien entity in the form of an attractive woman preys on men in *Under the Skin*. Both in recent terror attacks and in sci-fi narratives about the homegrown enemy, fortified cities and military installations can only do so much in the face of the threat from a relentless 'internal' Other. Being ontologically distant from the 'we' and not belonging to the same civilisation, the 'internal' Other cannot be assimilated: they are a non-human or de-humanised *monster* (as some in the news media have described European homegrown terrorists) that cannot participate in the human organisation of space (or the civilised, liberal, secular, human space) and therefore destroys it.

This widespread interpretation of terrorism as the marker of an inferior civilisation is shared by Slavoj Žižek (2015), among others. Referring to the 2015 Paris attacks, he argues that

> the problem with [terrorist] fundamentalists is not that we consider them inferior to us, but, rather, that *they themselves* secretly consider themselves inferior. This is why our condescending politically correct assurances that we feel no superiority towards them only makes them more furious and feeds their resentment.

as if implying that they lack something that would make them *real humans*. Liz Fekete (2009) highlights that the December 2001 *EU Common Positions* and *Framework Decision on combating terrorism* adopted a similar stance, viewing international terrorism, or 'the many-headed monster' in Tony Blair's words, as more dangerous for European values and way of life in its domestic dimension, and any 'form of support, active or passive' (p. 45) given to international terrorism as an act of terror itself.

In my view, representations of 'external' threats or 'internal' risks posed by allegedly dangerous homegrown *aliens*, or even migrant and refugee invasions, all share the same basic discursive structures. Cases in point are EU member states' violations of the non-refoulement principle, a central tenet of the Dublin Convention, and the news media's responsibility in associating the alleged 'invasion' with *moral panic* (Hall 1996; Chouliaraki and Zaborowski 2017; Bhatia et al. 2018; Pogliano 2019; Giuliani et al. 2020a). In both cases, colonial 'figures of race' are mobilised for the purpose of *Othering*, *distancing* and *monstrifying* the threat, thus naturalising the association of migrants, refugees and citizens from a postcolonial background with potential danger. The figure of deceiving, fanatic, and woman-hating Muslim male at the core of both Islamophobia (Sayyid and Vakil 2010, p. 3) and xeno-racism comes from a colonial archive that has been shared across imperial borders. So does the figure of the inherently violent black man posing a sexual threat to white women that Hartman (1997), Saucier and Woods (2014), and Danewid (2017) have traced back to Negrophobia and the slavery archive. Deployed in the past to legitimise the biopolitics of forced mobility and immobility within the space of the colony and the metropole, these figures today underpin the operation of the border system.

Although I acknowledge that Sivanandan's 'xeno-racism' is an effective concept for describing racism against migrants, asylum seekers and refugees, I disagree with the idea of tracing racist materials back to Europe's history and the Second World War exclusively (Fekete 2009, p. 42). Neither do I agree with the argument that xeno-racism is 'irrespective of colour' (Sivanandan 2001, pp. 1–5; Fekete 2009, p. 43): it is, in fact, fully aware of colours and sees them well through the lens of racist practices that have crossed colonial spaces since early Modernity. Finally, unlike Sivanandan and Fekete, I believe that xeno-racism aims not at excluding 'the new poor' but rather at differentially including them (Mezzadra and Neilson 2013) and turning them into exploited labourers and consumers while rendering their life precarious and perpetuating the privileges of hegemonic whiteness. In separating and connecting, excluding and including at the same time, the border sutures and distances histories, people and geographies, assembling meanings of *monstrosity* from cultural (national) and colonial archives (the figures of race) in order to legitimise itself, its bio- and necropolitics, and the system of exploitation it upholds.

Monstrosity may thus appear as an 'empty signifier' or a 'sign without meaning' as a result of the border's operations (Nuzzo 2019, p. 33; Nuzzo 2018). *Race* itself has been defined as an empty signifier (Hall 1997), but although both race and the border may look like containers filled with signs, neither is. No signifier is empty, as all signifiers arise from semiotic landscapes filled with historically stratified meanings. Building on the work of Stuart Hall (1997) and Colette Guillaumin (1972), I have defined race as a floating signifier of situated colonial, settler colonial, and postcolonial power relations (Giuliani 2019, p. 8). Race is a relational construct, never fixed once and for all but historically signified according to the symbolic material shaping specific semiotic landscapes and the power relations it serves. Race as a discourse interprets and modifies the body, symbolically and materially, through a variety of everyday biopolitical practices. Both the border and race emerged as the result of assemblages of pre-existing meanings.

In colonial modernity, the 'elsewhere' had to do with the *colonial* organisation of space. In our postcolonial postmodernity, time and space are fractured by the so-called multiplication of borders resulting from the multiplication of capitalist microsystems of surplus-value extraction and labour exploitation (Mezzadra and Neilson 2013). Now as then, borders and the 'figures of race' serve as a means of re-establishing the distance between people and between places. Today the racialised figures of threatening migrant and alleged terrorist are mobilised both physically and symbolically through the semiotic action of the border – not without resistance and resilience on the part of migrants themselves. What results is a circular relation between sign (the reproduction of postcolonial racism) and material effects (the border as a postcolonial bio- and semiotic technology), that is, between constructions of monstrosity engendering *moral panic*, the regulation of mobility, and the reproduction of translocal gendered and racialised power relations and inequalities.

Michel Agier (2018) has argued that borders connect and disconnect while walls separate, responding to a need for identity and identification that seems to be felt by both sides. But I agree with Wendy Brown (2010) that they too connect and disconnect – along with camps and colonies, I would add – because walls can be scaled and the space they delimit can be trespassed upon. Unlike Brown, however, who sees wall-building as a symptom of the loss of State's sovereignty over its territory and the search for (national) identity in post-Westphalian times (p. 116), I see walls as having the same functions as borders and the discourse on borders. My aim, unlike Agier, is to investigate the experience of the border on the side of the 'legitimate' and 'civilised' for whom protection is sought from monsters and from the 'illegitimate and uncivilised persons whose suspicious patterns of behaviour are to be targeted and apprehended' (Amoore and De Goede 2008, pp. 5–6). The use of walls and borders as biotechnologies can be explained, materially and symbolically, by the identity crisis facing Europe as well as the United States, Australia and Israel:

> The call for states to close and secure national borders is fueled by populations anxious about everything from their physical security and economic well-being to their psychic sense of 'I' and 'we'. Today, xenophobia is so overdetermined by the economic and political insecurities generated by globalization that even politicians cognizant of the limited efficacy of border fortifications lack discursive points of entry for discussing them.
>
> (Brown 2010, p. 69)

Building on this insight and on Brown's analysis of walled borders, and applying Fekete's analysis of the 'deportation machine' and 'suitable enemy' (2009, pp. 135–155) to the context of border management in the Mediterranean (Tazzioli and Garelli 2018; Fassin 2011), my aim here is to offer a postcolonial perspective on the crisis, focusing on a critical analysis of the racialised 'I' and 'we'. The lack of discursive points of entry mentioned by Brown can be explained by politicians' caution in addressing fears of racial conflicts and, ultimately, racial replacement. In this frame, Europe's identity crisis is to be understood as a crisis of white identity and white privilege generating *white anxiety* (for a broader reflection on *white anxiety* in the West, see Hage 1998, 2003; Yancy 2012). In my analysis of the semiotic, race-making function of the border, the important questions are: How is white privilege questioned?; How does the coming *alien* trigger *white anxiety* in Europe?

White anxiety in Europe can be traced back to two key issues: first, the porosity of its external borders, allowing entry to embodied witnesses of colonial and postcolonial violence and horrors; second, the failure of multiculturalism and assimilation policies to understand and contain postcolonial social, racial and gendered conflicts (which today are seen as inextricably tied to radicalisation and terror attacks) as well as the returning myth of Europe's homogeneous white identity and the need to preserve it. As I will discuss with reference to the imaginary behind DNA refraction in *Annihilation*, fears of contamination leading to an irreparable mutation of both individual bodies and the body of the nation (and, ultimately, to the extinction of the white/human race) lie at the root of racial anxiety towards migrants and national *aliens*.

Several esteemed colleagues have suggested I trace the imaginary of alien invasion to the precarity of Western ideas of technological and scientific progress, and indeed aliens have always been thought of as more advanced than humans, while Western humans, in particular, have always harboured doubts about their own supposedly superior progress. As Wells clarifies in the quote in Section 1 of this book, just as 'we' have colonised and annihilated the 'less civilised', there will always be the possibility that someone who is 'more civilised' may do the same to us. Rupert Wyatt's *Captive State* (2019), in which humans have been colonised by aliens who went on to become governors and 'Legislators', exemplifies this narrative. While I agree with this interpretation, I see it as complementing rather than opposing mine: in fact, if we posit that human 'doubts' today concern the deadly effects of the Anthropocene and warfare mentality, it follows that fantasies of alien superiority may have to do with the idea that something went wrong with Western progress, which could eventually lead to the extinction of the Anthropocenic Anthropos. The alien may be seen as the allegory of two different perceptions: that the 'wretched of the Earth' coming from the colonial past and postcolonial present is ready to retaliate, and that aliens are bringing with them a different outlook on the relation between humans and the Earth. The latter, as we will see in Section 3, embodies a critical conception of the relation between human Life, non-human Life and Nonlife that opposes Western foundational geontological and biopolitical dichotomies of we/they, nature/culture, global/planetary, and political/non-political. Embracing practices and concepts from the knowledge system of the outcast, the fugitive and the wretched of the Earth, aliens seem to present their alternative epistemologies as the last chance to prevent total catastrophe (Tsing 2015; Haraway 2016; Povinelli 2016).

The use of alien invasions as an allegory for the return of the barbarian is rooted in the fear of loss of identity that in Europe has seen nationalists, identitarians and white supremacists rally around the concept of 'great replacement', first popularised by the French nationalist Renaud Camus.[7] This theory, which sees the return of the barbarian as a deadly omen for white and Western civilisation(s), has become popular with an audience that goes well beyond European far-right militants and sympathisers, as demonstrated by research at the intersection of the security discourse, representations of migrants and refugees, and news media constructions of the terror threat. The semiotic landscape sustaining fears of invasion and disaster is made of widely circulated narratives such as those that made front-page in British mainstream news media in 2015: the April shipwrecks in the Mediterranean, the terror attacks on France, and the sexual assaults in Cologne on New Year's Eve,

> the use of simplified categories of origin (Africans, Arabs), the explicit reference to religion to characterise Others (Muslims), the analysis of cultural differences as inferiority and as an inherent source of conflict,

> and the resort to metaphors that imply invasions or savagery are examples of this (Holmes and Castañeda 2016). As explained by Chouliaraki and Stolic (2017, p. 8), 'Visualities of threat rely on the racialisation of refugees, where skin colour and clothing operate as signifiers of evil 'otherness' in 'our' midst – also reflected in animalistic references to 'swarms', 'flocks', or 'cockroaches' in the U.K. media.
>
> (Santos et al. 2018, p. 457)

As Júlia Garraio highlights referring to the mainstream news media's coverage of immigration to Germany,

> in Fall 2015, the anti-Merkel narrative *Wir schaffen es nicht* [We can't do it] – claiming that the arrival of migrants was overburdening Germany, the state was losing control and chaos was about to erupt – became dominant. This narrative was decisive in framing the coverage of the 2015/2016 New Year's Eve sexual assaults in Cologne and the rape-murder of a student in Freiburg in 2016 (Hermann 2019, 151) [...] In a context of ethnic anxieties and nativist backlash against immigration, the 'ethnicization of sexism' (Dietze 2016) which pervaded the mediatization of events in Cologne strengthened the public association between Islam and 'gender violence, feminism, and gender mainstreaming', thus reinforcing the positioning of 'Islam as the racialized Other to Germany, to Europe, and to the West' (Weber 2016). In sum, the sexual assaults were translated into already existing cultural interpretation patterns (Werthschulte 2017) that sustained the reading of the events as an attack on the German nation, which in turn strengthened the discursive positions of the far-right (Boulila and Carri 2017, p. 288–291).
>
> (Garraio in Giuliani et al. 2020a)

Fears of invasion are also constantly mobilised in Italian, Austrian, Hungarian, Spanish, Greek, Belgian and even Portuguese news media (Bhatia et al. 2018; Chouliaraki and Zaborowski 2017; Dixon et al. 2018; Santos et al. 2019; Torkington and Ribeiro 2019) when evoking Europe's Christian roots and identity, a discourse that has been referred to as 'cultural fundamentalism' (Stolcke 1995) or identitarian populism – both linked in Bruno Latour's opinion to the New Climate Regime (2018, p. 13–14). It could be argued, then, that the far-right and the identitarians are saying what they are expected to, openly stating that migrants lack bio-legitimacy, and that nations need to be 'monoracial' (Boeyink 2019, p. 63), 'monocultural' and 'monolingual'. But their views about increasingly weak external geopolitical borders and the loosening of 'national' bounds and identity are also shared under a different guise by cross-party pessimistic discourses on globalisation (e.g. by U.K. Labour Party leader Jeremy Corbyn and by former Italian prime minister and former Democratic Party member Matteo Renzi). Fear of invasion, which is clearly *fear of racial replacement*, is portrayed allegorically in films such as *War of the Worlds*, *District 9* and *Annihilation*, where non-humanness is a metaphor for racial and racialised cultural differences. There is no room for coexistence, nor for mixing or communication in far-right ideologies and white supremacist cultural products; they are simply deemed impossible lest the door to annihilation be left wide open. In the next chapter, I will explore how alien narratives conceptualise and present these issues and the symbolic material that shapes fears of invasion in the European imaginary.

5 Segregating the monster, eliminating the Otherworldly

Science fiction has long been regarded as a product of far-right or even neo-fascist cultures for indulging white or Western (neo-colonial) supremacy and genocidal fantasies against both the external enemy and the internal other. On the contrary, although imaginaries of extermination are sometimes employed, the greater part of science fiction not only uses the ultimate threat as a device for critically reflecting on human societies and affects, but also identifies post- or non-human Other(ed.) entities as peers with whom constructive relationships can be built. While there is merit in the criticism of films like *War of the Worlds* as ultra-conservative, Amoore and De Goede's dichotomy seems somehow outdated when it comes to new productions exploring the potential for coexistence between human, non-human and post-human beings, where the issue of internal other/external enemy is seen in new light.

For the purpose of my analysis, I have grouped film narratives into three different categories according to the specific biopolitical regime dealing with the *strange encounter*: segregation and elimination (*War of the Worlds*, *The Mist* and *District 9*), coexistence and empathy (*Monsters* and *Under the Skin*), and unsettling communication between humans and non-humans (*Arrival* and *Annihilation*).

Below, I will focus on filmic representations of segregation and elimination, exploring connections with the colonial archive of the colonised/enslaved and the constraints placed on their body and mobility. The Other is depicted as a non-assimilable (alien) and apparently invincible terrorist both in *War of the Worlds*, a remake of Wells' seminal classic, and in *The Mist*. The segregated and racialised immigrant in South Africa's *District 9* is instead seen as less 'Other', less 'foreign', for they are capable of coexisting with humans and turning them into a hybrid of both species. Dystopian narratives entangled with utopian future visions reproduce the discourse of the new social order envisioned by global and local governments, while rearranging notions of citizenships and territoriality to the point of including inhabitants of other planets and universes.

District 9's protagonist, Wikus, is a middle-level manager at MNU (Multi-National United), a private corporation tasked with evicting thousands of resident aliens from a slum in Johannesburg, South Africa, slated for gentrification. The story recalls the events surrounding the apartheid-era forced removal of 'coloured' people from District 6 and Sophiatown, where they had lived for decades, in a reinterpretation of new divisions and new Others living on the margins of South African society, where they are caught between organised crime and gentrification. In the words of Pier Paolo Frassinelli, *District 9* 'uses the mediation of science fiction genre norms to present a dystopian narrative that is not situated in the future, but in a present and recognisable place' (2013).

In the film, aliens have been confined to the District 9 slum after landing in South Africa a few years earlier and getting stranded there because of a spaceship malfunctioning. Unable to negotiate with the government as equals, they have been treated as if they were invaders, much like the migrants from Mozambique or Nigeria who move to South Africa in search of work and better living conditions, with whom they share the status of outsiders, or better, 'alien residents' (Frassinelli 2013) living outside the legal boundaries of South African citizenship. The film was shot in the area of Soweto that was the site of violent xenophobic attacks on migrants from the poorest neighbouring African countries in 2008. Since the mid-1990s the South African constitutional concept of citizenship has indeed been marked by ethnically exclusive features in the name of a supposed superiority over other African peoples.

After accidentally spraying himself with alien fluid, Wikus begins mutating into an alien or 'prawn', as the aliens are called because of their resemblance to the insects (*Libanasidus vittatus*) infesting the rubbish dumps of Johannesburg. As a result, he finds himself empathising with the condition of the migrants and subaltern, and is soon 'eliminated' by his boss, who is also his father-in-law, who considers him inept and spineless. Wikus finally crosses 'the boundary' when he offers to help an alien scientist who is trying to fix a device that will allow him to restart the spaceship. This device only works with alien DNA, the same DNA that, because of contamination during an eviction, has infected Wikus dooming him to mutation.

Every reference in the film, from descriptions of urban spaces to the race relations among humans and between humans and anthropomorphic aliens, seems to be an allegory of present-day South Africa. And yet, as Frassinelli (2013) points out, while *District 9* may be a harsh (and funny) critique of old and new forms of segregation and subordination in a society dominated by gender-based violence, racism and inhuman inequality, its stereotyped representation of the Nigerians reproduces tropes of the new essentialised Other in post-apartheid South Africa.

The film depicts the Nigerians as ruthless criminals, virile to the point of brutality, trafficking in guns, drugs, women and dirty money and exploiting the apparent 'innocence' of the aliens. On the contrary, what makes Wikus likeable is his vulnerability to his father-in-law and to alien DNA – which turns him into a mutant alien deprived of citizen status. He becomes a biopolitical object to be either left to its own destiny or mobilised to serve the interests of racist violence and class division. Audiences feel for him, because it is clear from the first scenes that he is the weak link in the chain of command and reproduction of white privilege: he will get his hands dirty, but he is more akin to an unwilling executioner than a hyper-masculine Terminator. The aliens' anthropomorphic features are not assigned randomly: Wikus goes from being 'not man enough' to being 'not quite human' (not entirely white), which is somehow equivalent in status in the triple ontology of masculinity, whiteness and South Africanness still dominant in contemporary South Africa.

While *District 9* is an allegory of apartheid and spatial segregation, the main theme of *The Mist* (2007) is the colony – the modern biopolitical *dispositif*. Frank Darabont's film is based on Stephen King's 1980 novel by the same name, also adapted into a TV series directed by Christian Torpe in 2017 (one season, ten episodes). In the film, after a violent storm hits the town of Bridgton, Maine, David Drayton (Thomas Jane) leaves home to buy supplies with his eight-year-old son, Billy, and a neighbour. Along with other town residents, they will find themselves trapped in the local supermarket, which becomes a sort of walled colony under attack by horrific monsters hidden in a thick mist. The alien invasion, as the young soldier Wayne Jessup reveals, is the unwanted outcome of the Arrowhead Project, a military experiment gone awry: aliens began flowing in when the gates between the known universe and parallel dimensions were opened. Jessup is eventually made a scapegoat for the accidental arrival of the aliens and expelled by the community. Once outside, he is immediately killed by the monsters. Meanwhile, race-, class- and gender-based conflicts arise in the small community where Mrs Carmody (Marcia Gay Harden), a religious fanatic who believes that Armageddon is near, successfully manages to proselytise several survivors – even more so after she is spared by giant insects attacking the supermarket. David and his son, along with others in the community who are sceptical about her end-of-the world predictions, try to set up a collective defence system and find a way to escape the supermarket.

As an allegory of the colonial formations that originated American society, the trope of the store-colony not only recalls a foundational history of violence but, as in *The Walking Dead* and *World War Z*, is still imagined as the place where life and society can be rebuilt and from where civilisation can spread. It obviously recalls the famous mall in George Romero's *Dawn of the Dead* (1978) – the popular destination for American society's ritualised shopping and recreation, the place where the community traditionally gathered and socialised in the golden age of consumerism. The group of unbelievers comprises a retired school-teacher, Irene, an elderly man, Dan, Bobby's teacher Amanda, and Ollie, the supermarket manager. They are the wise outsiders who do not buy into biblical apocalyptic prophecies or condone violence in the panicked community, the enlightened ones who are not willing to repeat a history of violence nor be the victims of fear, rage or hatred. After the attack from the giant insects gains Mrs Carmody an even bigger following, Amanda alone still has faith in humanity and civilisation. David argues that civilisation matters but only

> as long as machines are working, and you can dial 911. But you take those things away, you throw people in the dark and you scare the shit out of them, no more rules, you will see how primitive they can get.

Dan's and Ollie's outlook about the future is even gloomier: the former believes that 'if you scare people badly enough, you get them to do anything, they'll turn to whoever promises a solution', and the latter is persuaded that 'as a species we're fundamentally insane. Put more than two of us in a room, we pick sides and start dreaming up reasons to kill one another. Why do you think we invented politics and religion?'.

The following day, they decide to leave the colony after hearing Mrs Carmody incite the mob to sacrifice Billy to the monsters, but they soon get stranded with an empty tank in a deserted area infested with aliens. Rather than face a horrible death, they agree that David will shoot them one by one with Ollie's gun and then take his own life. But after killing them, David is left with no bullets and exits the car, only to discover that the mist is disappearing and the army is on their way to rescue the survivors. The useless sacrifice of the wiser residents of the colony may be seen as a metaphor for the impossibility of escaping Western bio- and necropolitics and their violence in the form of camps, colonies and colonial (genocidal) wars. In the last scene, David catches a glimpse of the fearless woman who had left the store-colony alone, ignoring naysayers, to go find her children: she is perched on top of an army tank with other survivors.

Steven Spielberg's *War of the Worlds* (2005) is the third and last film in this group. Adapted from Herbert G. Wells' 1898 novel, the film, just like Byron Haskin's 1953 version and *The Mist*, features a deadly confrontation with the unexpected alien invaders, a sort of *mors tua vita mea* on a global scale opposing humankind and alien-kind. The protagonist, Ray Ferrier (Tom Cruise), is a white, working-class divorcee from a largely non-white suburb in New Jersey. Despite being not too involved in the new affluent life his children have started with their mother and her new partner, he is the one who successfully manages to rescue them and bring them home – in what has all the appearance of a vindication of the working-class white docker who spends his days working with heavy machinery but also clearly knows how to keep his loved ones safe. His little girl Rachel, in particular, suffers from panic attacks and seems unable to survive on her own. Ray's good intentions and solidarity with other human beings are contrasted in the film with the selfishness of the mob that attacks the vehicle carrying the

Ferrier family. Still, there are other 'good' people out there, such as those who, finding Rachel alone, offer to help her and take her with them.

The enemy seems invincible: monsters drive huge tripod machines, with screens on which human bullets and bombs uselessly crash and laser beams that incinerate targets in seconds. Their arrival seems motivated by an urge to destroy, until it becomes apparent that the purpose of human abduction is to collect blood and use it as fertiliser for alien vegetation. Yet, what they hoped would be their last chance for survival leads instead to their destruction: they have no immunity to the bacteria residing in the human body. Earth's organic life, rather than military technology, is what ultimately defeats them as they fall to the ground one by one. Man always wins out in the end, and humankind's steady progress emerges unscathed from the encounter/conflict with the Other(s) who come(s) from outer space. The aliens, whose spacecraft has apparently been hidden underground since before man can remember, did not consider the effects that contact with human bacteria would have on newcomers.

The alien is, therefore, the absolute evil, biologically incompatible and far removed from the members of the moral community and their shared beliefs. A happy ending is provided by the reconstitution of the Ferrier family's moral and affective community. Finally reunited in Boston at Ray's in-laws, they seem to have overcome past class and generational conflicts, especially as regards Ray's relationship with his son, Robbie, who in the early scenes seems to regard his father's masculinity and (lower) breadwinner status as below par. They are the typical bourgeois, heteronormative family, whose alien member (Ray himself) eventually leaves to follow his own path. No segregated permanence or actual contamination is allowed with the alien element who is, in fact, biologically incompatible and phenotypically unrelated to humans.

The relation with the *distanced alien-ed* Other is paradigmatic of an impossible intimacy in these films. In *The Mist*, the colony and its rules seem to be the only chance for survival against the absolute enemy, although here the survivors must submit to extreme bio- and necropolitics dictated by fanaticism and ignorance Only two options are available to those who leave it: they will either get killed by the monsters or rescued by the military – a symbol of order and civilisation. Where establishing a self-protecting colony is not feasible and the absolute enemy is unrestrainable, suicide is the only chance to 'survive'. In *War of the Worlds*, aliens have been hiding underground for millions of years, observing humans and preparing a genocidal attack. Their attempt at using human blood as fertiliser is unsuccessful: the two species are too incompatible for *miscegenation* to occur. A certain degree of compatibility is instead seen as a possibility in *District 9*, where the hero undergoes a non-destructive transformation; he will, however, be segregated by those who wish to deny any possibility of coexistence with Otherness. Communication between the two species can occur, but only within the very narrow modes and walled spaces of a segregated society where aliens and mutating humans are treated as Mamdani's *subjects*.

6 Co-existing in walled spaces

There are some similarities between the alien invasion in *The Mist* and that of Gareth Edwards' *Monsters* (2010) mistakenly landing on Earth near the U.S.–Mexico border. Both *reveal* the powerlessness of borders and walls in the face of an overwhelming entity that enters the space of the 'we' through a wide open door or is already there, thus signalling vulnerability rather than strength. As Wendy Brown maintains:

> [I]t is the weakening of state sovereignty, and more precisely, the detachment of sovereignty from the nation-state, that is generating much of the frenzy of nation-state wall building today. Rather than resurgent expressions of nation-state sovereignty, the new walls are icons of its erosion. While they may appear as hyperbolic tokens of such sovereignty, like all hyperbole, they reveal a tremulousness, vulnerability, dubiousness, or instability at the core of what they aim to express – qualities that are themselves antithetical to sovereignty and thus elements of its undoing.
>
> (2010, p. 24)

In *Monsters*, aliens hatch from spores in trees, then keep growing in the rivers of the Mexican forest, and finally return to the ether as giant octopuses the size of a five-floor building. They only react when provoked, especially by the U.S. infantry, armoured vehicles and air force, and retaliate against humans regardless of their age, gender, class and colour. Andrew (Scoot McNairy), a war journalist, is hired by an American tycoon to bring his daughter Samantha (Whitney Able) back from Mexico, where she is doing volunteer work with the residents of the contaminated area, to her fiancé in the United States. The film touches on many themes: war photography, volunteer work, the relationship between nature and war, and the boundaries marked by citizenship, with the white affluent rich on one side and the non-white alien and Mexican poor on the other. The setting particularly lends itself to describing the alien condition in a way that is both harsh and delicate.

The border area surrounding the wall between the U.S. and Mexico is one of the bloodiest of the world (or used to be, before migrants started using the Balkan Route to reach Europe). In the last 25 years, it has divided, killed, segregated and allowed the exploitation of low-wage, non-unionised labour force at *maquiladoras* established by U.S. multinationals in Northern Mexico's border cities. The *indocumentados* who try to cross the border face deportation and risk dying in shootouts with U.S. border patrol agents or being held at gunpoint by armed vigilantes. As if anyone 'invading' U.S. territory from the South were a bloodthirsty alien.

As Wendy Brown reminds us, walled borders are a global phenomenon:

> Some walls are little more than crude fences through fields, while others are mammoth, imposing structures heavily adorned with contemporary surveillance technology. They would also seem to address a diverse array of problems. Most South Asian nation-state walls, for example, target immigrants, while most of the walls of the Middle East are built in the name of security from terrorism. The Uzbekistan wall against Kyrgyzstan was prompted by border conflicts, while the Ceuta and Melilla walls in Morocco seek to prevent these Spanish enclaves from becoming staging grounds for Asians and Africans seeking to get to Europe. The berm across Morocco's Western Sahara aims to appropriate disputed territory and some also regard the Israeli wall as a land-grab.
>
> (2010, pp. 27–28)

But walls are as porous as the borders themselves, as evident both in this film and in *WWZ*, where a wall separates Israel from Palestine and humans from the rabid monsters. In fact, although 'compared with the evanescent, protean, and depthless traits of late modern culture and politics, walls seem solid and permanent, and appear to lack capacities for guile and dissimulation' (Brown 2010, p. 80), the protagonists trespass the red zone with the help of migrant smugglers who know how to escape border patrol detection. As with all walled borders, bio- and necropolitics along the U.S.–Mexico border select and filter people, as well as enhancing the differential ontology in which migrants' subalternity and illegal-ised mobility are grounded. Alien monsters, too, seem capable of circumventing biopolitical and necropolitical *dispositifs* and annihilating the military apparatus. Walls, like colonies in *The Mist* or segregated slums in *District 9*, appear for what they are in post-apocalyptic narratives: rather than providing an effective barrier against the threat, they are mere performances of a modern and colonial idea of a segmented space that is constantly overthrown by restless insubordinate people exerting their 'right to escape' (Mezzadra 2001, 2004).

Monsters, like migrants, are seen here in search of an environment where they can settle. Without denying the importance of the 'objective causes' at the root of contemporary migrations, that is 'wars and misery, environmental catastrophes and political and social tyrannies prevailing in vast areas of the planet' (Mezzadra 2004, p. 270), both, individually as well as collectively, make their way across the walled border and the red zone in search of what is imagined as 'a better life'.

> for migrations to exist, there must be an individual motion (made concretely by a concrete woman or man, embedded in family and social 'networks', but nonetheless capable of *agency*) of desertion from the field where those 'objective causes' operate, a reclaiming precisely of a

> 'right to escape', which, even if most of the time unconsciously, constitutes a material critique of the international division of labour and marks profoundly the subjectivity of the migrant also un the country where she/he chooses to settle down.
>
> (Mezzadra 2004, p. 270)

As I have argued in this book, the critique instantiated by migrants today materialises in their opposition to postcolonial *dispositifs* of *bordering, distancing* and *Other(world)ing* that relegate them to the physical and symbolic space of the unworthy, the inferior, the dangerous – that is, allegorical aliens or monsters. It is by defining them as such that they are made into exploitable subalterns, dispensable lives, or killable individuals (Butler 2009; Asad 2007; Mbembe 2003; Sousa Santos 2018). The giant octopuses breaking through the ever-expanding defence barriers and military technology in the film's fictional world – and emerging unscathed – recall not only the permeability of human segregated geographies but also the discursive and semiotic features of absolute Otherness and illegal border crossing.

But monsters here are also the allegory of a *geography* of disaster that from a confined space, a parallel dimension – the elsewhere – has spilled into the so-called 'civilised world', as Amanda calls it in *The Mist*. Alien monsters here and in general are the allegory of the apocalypse caused by the absence of a frontier between the world and otherworldliness, order and chaos, technology and nature, human and non-human life. For that reason, then, there is usually no sympathy for the non-human nature capable of destroying civilisation. *District 9*, *Monsters*, *Arrival* and *Annihilation* are all narratives in which humans willingly or unwillingly need to come to terms with mutating aliens, human mutants and the permanent 'mutation' of the *world as the 'we' knew it*.

In *Monsters*, the alien condition and the fears it triggers are seen through the eyes of two white affluent young people, Andrew and Samantha, who are, nevertheless, meant to make audiences empathise with the unfortunate plight of the Other (e.g. when they are forced to make their way through the contaminated area with the same makeshift equipment used by migrants and the local poor to cross the border into the American Eldorado). The film takes a delicate approach to the representation of human suffering and solidarity between those who live in the borderlands, and even to the depiction of aliens, as evident from the scene in which Andrew and Samantha find shelter with a Mexican family and share, albeit temporarily, their living conditions, or when the remains of a family decimated by the giant octopuses are found in the forest. Andrew covers a little girl's body with his jacket as he reveals to his traveling companion that the photo of a child in distress is far more profitable than that of a smiling one.

As the two protagonists finally, and predictably, exchange their first coveted kiss at a gasoline pump, they see two giant octopuses move towards each other, moaning and exchanging some kind of tentacular contact as if they were mimicking them. This scene, set in a deserted New Mexico town where the only resident left is an old lady who drags her belongings around in a shopping cart, is full of meaning. What makes it unique, both from a symbolic point of view and cinematically, is that it makes us look at the giant octopus 'Others' fondly, as if they were our own pets – the difference being that these creatures, unlike our pets, are huge, threatening, and we cannot control them, as they are 'not yet classified by science and war intelligence' (Foucault 1961). Aliens are not humanised, quite the contrary, but the audience sees them through the friendly eyes of the protagonists, thus including them within the boundaries of enlarged citizenship, as if the scene translated into visual language what Donna Haraway would call a 'non-speciesist' notion of citizenship – that is, not restricted to humans and breaking down the barriers between so-called species (2003, 2007). In *Monsters*, alien forms germinate in peaceful coexistence with nature, just like the bacteria from our skin and body fluids. This similarity they share with living things makes them easy to accommodate in 'our' world.

Both *Monsters* and *Arrival* offer what appears as an anti-Anthropocene or post-human philosophical response to the xeno-racism described by Fekete and Sivanandan: in a world in which symbolic and geographical divisions allowed Western power relations to be built and maintained through racial genocide and labour exploitation, changes to that *geography* bring chaos. Where bio- and necropolitics of elimination, appropriation and surplus value extraction (Dussel 1995; Svampa 2019; Cole 2000) that regulated all forms of life to allow for profit maximisation and the preservation of racial privilege are no longer viable, the alien's *risky* body poses a major *risk* to society, radically disrupting capitalist production and the parameters defining life in the West (such as duration, reproduction and health). Unpredictability enters the picture and disrupts the human order; nature erases the vestiges of Cartesian thought placing humans at the centre of the scene and shielding them from the effects of intra-active phenomena (Barad 2008). Unlike the appropriation and diminishing of the Other's suffering that I discussed in Section 1, empathy for the monster here denotes awareness of their quest for well-being and a honest but hopeful awareness of a shared condition of vulnerability and precariousness: all humans are *bodies at risk* in a new world where rationality and especially capital's rationality are made almost irrelevant.

It is this awareness, coupled with pity and compassion, that prevents Laura (Scarlett Johansson) from killing her latest prey, a young man with neurofibromatosis, in Jonathan Glazer's *Under the Skin* (2013). In the film, loosely based on Michel Faber's 2000 novel by the same name, Laura is an alien who for reasons that are unclear is commissioned to provide a continuous supply of human flesh. To do so she takes the human appearance of a young woman who was raped and killed and then left on the side of a highway. After heartlessly killing the men she picks up while driving her van or in bars across Scotland, Laura is about to prey on her last victim when a sudden bout of compassion prevents her from carrying out her task. She lets him go and runs away, then abandons her van in the Scottish Highlands and is chased by a gang of aliens with the appearance of motor-bikers. A gradual process of awareness and self-discovery ensues, which reaches a point of no return when, after meeting a man who offers to shelter her, she discovers taste and sex, and finds out she has genitals. Shocked by her own physicality and emotions, she runs away and hides in a cabin in the forest, where she is eventually killed by a man who attempts to rape her but finds out she is not human.

In *Under the Skin*, the allegory of the predatory Other is projected onto male humans. What I find interesting is the fact that compassion, intimacy and empathy emerge when the Other (the human) is seen not only as a predator like all other men, including the one who eventually kills Laura, but also as a vulnerable being. It is because of the Other's vulnerability that she makes herself vulnerable, stops killing and succeeds in getting intimate with the Other. It is because of this proximity that she breaks all barriers and eventually loses herself. But in a (human) world of violence, getting out of the fence is dangerous: as soon as she is recognised as an internal enemy, she is immediately eliminated without being given a chance to explain herself. Her brief encounter with the man who offers to shelter her is her only chance at meaningful communication with the Other: no questions are asked; no explanation is expected. Language here is about silent *skin-to-skin* proximity and co-existence:

> It is through getting closer, rather than remaining at a distance, that the impossibility of pure proximity can be put to work, or made to work. [...] The multiple ears that are required to 'hear' the other, without transforming this other into 'the Other' or 'the stranger', are ears that are alive to, or touched by, the sensations of other skins. Such sensations open this other to other others, who are not simply absent or present in the *skin-to-skin* of the encounter. An ethics that keeps alive the circuit between mouths, ears and skin is hence not about making her body present. It is the act of getting closer to this other's skin that prevents us from fleshing out her body as 'the stranger's body'.
>
> (Ahmed 2000, pp. 157–158)

7 Unsettling communications

There is another possibility beside modern and colonial bio- and necropolitical *dispositifs* against the alien monster and utopias of co-existence and intimacy. Here, I will discuss unsettling communication, which is unsettling because it radically transforms *the world as the 'we' knew it*, erasing borders, boundaries and identities. Both in *Annihilation* and in *Arrival* the alien invasion and its unpredictable consequences are a metaphor for the encounter with the unknown 'out there' trespassing the binary border between human and alien and disrupting the geontology of Life and Nonlife. An effort is made in both films to imagine the encounter and its consequences as 'change' to which we, the humans, need to adjust (*Annihilation*), or as a 'change' that will increase human knowledge and understanding (*Arrival*). While communication in the former is against humans' will, acceptance from some in the latter allows the whole community to benefit from the exchange.

While *Monsters* seems to suggest that coexistence is possible, what complicates the issue in *Annihilation* is genetic modification, a sort of definitive *translation* of the world the 'we' used to know into a new order of things in which the environment and the human body are recombined into a mutated whole called the Shimmer, the alien-generated environment that escapes (human) predictability. Unlike *War of World*, *The Mist*, *District 9* and *Under the Skin*, where unilateral exploitation or extermination are possible, nothing can be done here to prevent nature from erasing the distance enforced by the technology/nature divide. The *alien-ed* nature of the Shimmer will expand 'until it encompasses everything', as the expedition's chief scientist, Dr Ventress (Jennifer Jason Leigh), predicts just before dying. By randomly refracting and recombining radio waves and DNA, the Shimmer makes human identity biologically indistinct from other living organisms and inanimate material, invalidating human capacity to control their behaviour, master nature, exploit resources, and (re)produce (the Anthropocene is over). DNA recombination and radio waves refraction do not necessarily have monstrous effects: according to the biologist Lena (Natalie Portman), they can create forms of life of extraordinary beauty, and the physicist Josie Radek (Tessa Thompson) argues that rather than getting to know it or fighting it, one should submit to it and let it bring about change.

Interestingly enough, the expedition team that Dr Ventress, an army psychologist, put together is composed of five women, four of whom are scientists. Beside Lena, the geomorphologist Cass Sheppard (Tuva Novotny) and the paramedic Anya Thorensen (Gina Rodriguez) are also members of the expedition team. They all have 'a past': Lena has lost her husband and Sheppard a daughter who died of leukaemia. Thorensen is a recovering alcoholic, and Radek used to self-harm. Dr Ventress' story is apparently a mystery, but Lena suspects she has terminal cancer. Except for Lena, who initially volunteers out of love for her husband, Kane (Oscar Isaac), the affective community that Dr Ventress has created is bound together by a self-destructive drive.

Lena's husband is not dead; in fact, he has just returned from a year-long mission to the Shimmer. Apparently, he is the only survivor from a previous army expedition whose destiny will become clear to Lena and the others when they find a memory card in the soldiers' barracks, with a video showing their genetic mutations. While Sheppard and Thorensen are torn to pieces by a mutant bear, Radek, who is herself undergoing a visible mutation, decides to stay in the Shimmer. Dr Ventress is the first to arrive to the lighthouse (that is where it all started, with an alien meteor crashing into the lighthouse in the opening scene). There, Lena discovers that the real Kane, unable to accept his mutation and duplication, has set himself on fire with a phosphorus grenade after asking his double to go find her.

Kane's double, who has reached Lena's home in the opening scenes, is on his death bed at the army base with multiple organ failure. After setting the lighthouse on fire and destroying the Shimmer, Lena returns to the army base, the only survivor from her expedition team. After being interrogated, she is put in isolation. In the meanwhile, Kane's double is out of danger and the last scene sees Lena and him hugging after confessing to each other that somehow both are products of the Shimmer.

The film's title refers to the destruction of the world as we know it: 'our bodies and minds will be fragmented in their smaller parts until no one part remains … annihilation', says Ventress just before dying, or rather, before mutating into a sort of genetic matrix blob. Towards the end of the film, Lena tells the officer who's interrogating her: 'I don't think it wanted anything'.

OFFICER: He came here for a reason, it was mutating our environment and destroying everything.
LENA: It wasn't destroying. It was changing everything. It was making something new.
OFFICER: Making what?
LENA: I don't know.

The film ends with the disappearance of the Shimmer. Nonetheless, its effects are visible on Lena's body. It is not known whether she and Kane's alien double will generate new refractions and mutations or the transformation will end with them, but in any case the world has been transformed

Figure 2.3 Still from *Annihilation* (2018), directed by Alex Garland. Lena (Natalie Portman) is reaching the lighthouse.

through unwanted communication, and new forms of life, geographies and ontologies have emerged.

While mutation in *Annihilation* is first of all biological, in *Arrival* it is related to the perception of time. In both cases, it is permanent: the invading Other is here to stay. Gone is the divided world and with it the solid epistemic, bio-physical, linguistic, geographic and political boundaries safeguarding humans' space. The alien presence cannot be erased: as in a reverse colonial invasion, the coloniser's physical (or epistemic) presence will gradually mutate and then put an end to the status quo (*Annihilation*) or contribute to improving it (*Arrival*). As discussed in the previous section with reference to monstrosity, fears of mutation lie at the foundation of Western culture and politics: they belong to a transnational colonial archive, at once local and transnational, providing the symbolic material that legitimises colonial expansion, extermination and the civilising mission as well as global wars (Giuliani, 2016b; Hage, 2016; Arata 1996; Asad 2007).[8] *Annihilation* deals with fears of miscegenation and bodily mutation and in a way introduces the arrival of the Chthulucene. Its beautifully recombined monstrosity can be seen either as the result of evil, barbarian invasions, punishment for human arrogance towards nature, or, on the contrary, as Haraway would maintain, as nature's attempt to minimise the ruinous effects of neoliberal subjectivity. In Chapter 6 of Section 3, *The Girl with All the Gifts* (2016) and the beauty of hybrid children's human and non-human life will allow me to further articulate my reflection on the Chthulucene and offer an alternative epistemology to the Anthropocene, grounded in the convergence of care, self-care and earth-care and in the interdependency of human Life, non-human Life and Nonlife.

Arrival deals with fears of loss of linear or predictable temporality. Forty-eight hours after the arrival of 12 alien spaceships, the U.S. professor of linguistics Louise Banks (Amy Adams) is contacted by Colonel G.T. Weber (Forest Whitaker) because of her previous work translating Iranian insurgents' videos from Farsi. Although disappointed at the U.S. intelligence's means and aims, she agrees to help and after reaching an army base in Montana, she and Ian Donnelly (Jeremy Renner), a physicist, are taken inside the spaceship to decipher the aliens' language and ask questions. Quotes from classic alien iconography – specifically from Herbert G. Wells' *War of the Worlds* and its film adaptations (1953 and 2005) – are seemingly a mark of Villeneuve's interpretation of the trans-worldly encounter. Here too, as in *Monsters* and in both adaptations of *War of the World*, aliens resemble giant octopuses with multiple tentacles (heptapods in Villeneuve's film, tripods in the other two). Yet, rather than being vertically kidnapped by alien spacecrafts like in Spielberg's film, scientists and soldiers are expected to voluntarily climb up a vertical tunnel where gravity is reduced.

As Louise and Ian start researching alien language and its complex circular symbols, she begins to have visions of a child who seems to be her own, a girl named Hannah who will die of a terminal illness as a teenager.

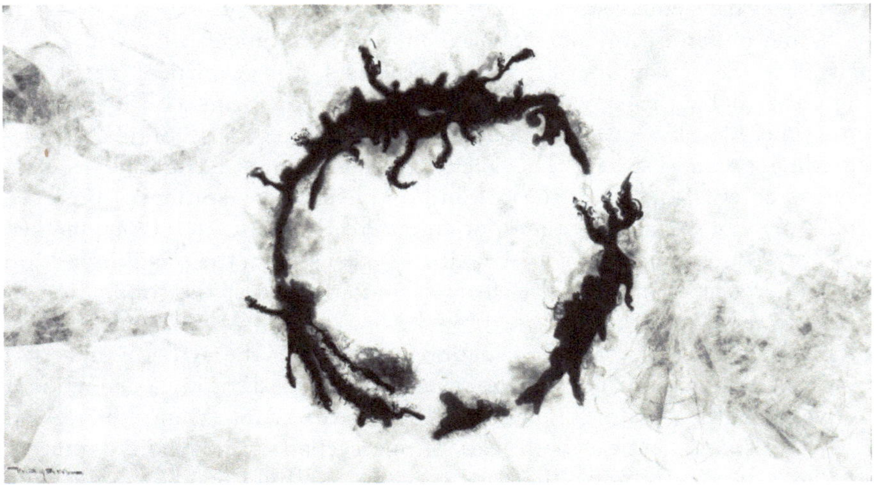

Figure 2.4 Poster from *Arrival* (2016), directed by Denis Villeneuve, showing alien ideograms.

When Louise has enough shared vocabulary to ask why the aliens have come, they answer 'Offer weapon'. However, China translates this as 'Use weapon' and promptly breaks off all communication. China, as Louise discovers, uses mah-jong tiles to interact with the aliens, turning every conversation into a matter of winning or losing. Other nations follow China's lead, and soldiers plant a bomb in the spaceship that has landed in Montana. Unaware, Louise and Ian re-enter the alien vessel, and the aliens give them an extremely complex message. Just before the bomb explodes, Ian and Louise are ejected from the spaceship and left unconscious on the ground. When they awaken, the military is preparing to evacuate, and the spaceship has moved out of reach.

Ian discovers that the symbol for time is present throughout the message, and that the writing occupies exactly one-twelfth of the space in which it is projected. Louise suggests that the full message may be split among the twelve vessels, implying that the aliens may want all countries to share what they learn.

China's General Shang issues an ultimatum to the aliens, demanding that they leave within 24 hours. Russia, Pakistan and Sudan follow suit. Louise goes alone to the spaceship, and a shuttle is sent down to transport her inside. One of the aliens explains that they have come to help humanity, for in three thousand years they will need humanity's help in return.

Louise returns to the camp as it is being evacuated and tells Weber that the alien language is the 'weapon', the 'tool', the 'gift' they are offering: learning someone else's language helps people change their way of thinking and adopt a new identity and, in particular, a different conception of time. As Sakai and Mezzadra highlight referring to translation, the latter is:

> at once a mechanism of domination and liberation, clarification and obfuscation, commerce and exploitation, opening up to the 'other' and appropriation. Translation, to further explicate its constitutive relation with the concept and institute of the border, produces both bridges and walls.
>
> (Mezzadra and Sakai 2014, p. 9)

But in this case translation is communication and as such carries a transformative power that can change the very essence of the 'we'. Aliens' perception of time is not linear: they will share this gift with humans, allowing them to experience 'memories' of events that have yet to happen. The very fact that Louise is learning the alien language explains her capacity to see her future with Hannah.

She has visions of a United Nations gala, where China's General Shang thanks her for persuading him to call off the attack by reaching him on his private number and reciting his dying wife's words: 'In war there are no winners, only widows'. Shang will know, once the whole humanity shares the 'gift', that she needed to make that phone call.

In the present, Louise steals a satellite phone, calls Shang and recites the words. Immediately after, the news announces that the Chinese will stand down. The other countries follow suit, and the 12 spaceships vanish. During the evacuation of the camp, Ian confesses he has feelings for Louise. She will agree to have a child with him despite knowing their fate: that Hannah will die and Ian will leave them, unable to cope with the revelation that she knew all along. But as she tells Hannah in the final scenes, as images of their family life with Ian appear on the screen: 'This is where your story begins, the day they departed. Despite knowing the journey and where it leads, I embrace it. And I welcome every moment of it'.

The kind of mutation envisioned in *Arrival* will enhance humans' ability to prevent bad things from happening and develop a more peaceful world of which aliens are rightful members. In fact, there is no immediate mutation in the bio-physical structure of human beings or the environment. The door is left open for new transformations that will offer a new experience of time, allowing humans to go back and forth and connect past, present and future. This does not imply a critique of human progress and its history of violence and exploitation. But calling for a more responsible approach to the world, one that acknowledges affects and *strange encounters* as the engine of humanity, paves the way to a critique of the ideals of universality, abstraction, neutrality and objectivity at the basis of the Anthropocene.

8 Conclusions

In this section I have focused on distanced/absolute Others as products of *Outsiding* practices operated through *geography* and targeting those regarded as a threat (Foucault 1961; Said 1978, p. 7). I have started my reflection with an analysis of the role of bio- and necropolitical *dispositifs* in the construction of monstrous human and non-human figures from ancient times to the birth of modern science fiction in the nineteenth century. In particular, I have explored the effects of such constructions on individual and group mobility across colonial borders and in the metropole in modern and postmodern times.

Through representations of the alien figure in recent sci-fi films I have explored practices of encampment, segregation, colony-building, wall-building and bordering in the past and the present and in dystopian futures. I have connected these *dispositifs* to a critique of the Anthropocene – the thinking behind the Anthropocene, its geontologies and bio- and necropolitics – and to the separation between Here and Out there, civilisation and barbarity, technology and nature, catastrophe and disaster, places *of* disaster and places *for* disaster, worthy and unworthy, worldly and otherworldly.

Alien Other(world)liness has helped me understand how these dichotomies have generated discourses of migration, terrorism and cultural or racial difference since 9/11. In fact, actual and symbolic distancing of contiguous geographies, cultures and people has produced, in the frame of the recent history of terror attacks and the crisis of multiculturalism, the idea of absolute Otherness – the ones who come from 'Out there'. I have drawn parallels between aliens – the allegory of the end of *the world as the 'we' knows it and wants it* – and migrants and terrorists, depicted as engendering annihilation, producing irreversible mutations and bringing catastrophe to the West. In this context, people's mobility appears as a social movement closing the distance and borders as the semiotic *dispositif* used to legitimise the *pretence* of an ontological fracture. Some of the films I have considered in my analysis propose a different reading of the so-called 'invasion' that builds on critiques of the Anthropocene – that is, of a conception of Earth as univocally transformed by human presence and action – and the Cartesian mentality that informs it. *Under the Skin*, *Monsters*, *Annihilation*, and *Arrival* offer a different perspective on the *strange encounter* that paves the way not only for future coexistence but also for interesting epistemological alternatives to a *war of the worlds*. Far removed from the epistemology of hybridity celebrated by some post-colonialists as an end to unequal postcolonial power relations (Ahmed 2000, p. 24), and a step forward from Zygmunt Bauman's idea that the issue is not 'how to get rid of strangers and the strange, but how to live with them – daily and permanently' (1997, p. 55), mutation allows us to rethink the elementary structures and operations of those same power relations in light of the Chthulucene.

The viewpoint of *Under the Skin*, where empathy is reserved for the vulnerable, is that the Other is a human male – violent and sexually aggressive. *Monsters* imagines a future in which aliens and humans can move beyond encamped geographies. *Annihilation* represents irreversible mutation as beautiful and, finally, *Arrival* challenges traditional conceptions of universality, abstraction, objectivity and neutrality of knowledge, language and time. In all these films, post-human embodied communication/translation between humans and aliens is the unavoidable destiny separating the present from the future and involving permanent mutation and a radical challenge to anthropocentrism and capitalism.

Notes

1 I will return to this topic later in this section and discuss it in greater detail in the third and last section, where I will tie together my reflections by looking at how the dichotomy we-here/them-out there characterising the human world and the planet has been constructed in *history* and *geography*.
2 I am grateful to my friends and colleagues Lars Jensen, Gianluca Bonaiuti and Marilena Indelicato for stressing the importance of my positioning towards my use of metanarratives.
3 Since ancient times, and especially during the French Revolution, kings, queens, court life, the clergy and the nobility have been also monstrified and described as vampires and molochs, as evident from Saint Just's political thought. See Foucault (1999).
4 Among the genre's milestones are the following U.S. productions: *The Day the Earth Stood Still* (1951), *The Thing from Another World* (1951), *It Came from Outer Space* (1953), *Invaders from Mars* (1953), *Invasion of the Body Snatchers* (1956), *Earth vs. the Flying Saucers* (1956), *Kronos* (1957), *I Married a Monster from Outer Space* (1958) and *Village of the Damned* (1960).
5 www.amnesty.org.uk/worlds-deadliest-sea-crossing-mediterranean.
6 Set in 1962, *The Shape of Water* is a steampunk fantasy film about a mute janitor at a high-security government laboratory who falls in love with a captive amphibious creature.
7 Namely: British National Party and UKIP, German Alternative für Deuschland and Pegida, the Italian Lega, Forza Nuova and Casa Pound, the French Front National, the Greek Golden Dawn, the Movement for a Better Hungary, Sweden Democrats, the Spanish Vox and Falangists, the Portuguese Chega and the Austrian Freedom Party.
8 For a critical analysis of the topos of the colonised land as a body to be penetrated, fertilised and mutated to make it fit for the newcomers, see, among many, the seminal work of Anne McClintock, Anja Loomba, Leela Ghandi (1998) and Ann Laura Stoler.

References

Aas, K. F. and Gundhus, H. (2015) 'Policing humanitarian borderlands: Frontex, Human Rights and the Precariousness of Life'. *British Journal of Criminology* 55 (1), pp. 1–18. https://doi.org/10.1093/bjc/azu086

Agamben, G. (1998) *Homo Sacer: Sovereign Power and Bare Life*. Stanford: Stanford University Press.

Agier, M. (2018) *The Jungle: Calais's Camps and Migrants*. Cambridge: Polity Press.

Ahmed, S. (2000) *Strange Encounters*. Abingdon and New York: Routledge.

Ahmed, S. (2004) *The Cultural Politics of Emotions*. Edinburgh: Edinburg University Press. http://doi.org/10.4324/9780203700372

Allen, T. W. (1994) *The Invention of the White Race*. London: Verso (this edition 2012).

Amoore, L. and De Goede, M. (eds) (2008) *Risk and the War on Terror*. New York: Routledge. https://doi.org/10.4324/9780203927700

Andersson, R. (2018) 'Profits and predation in the human bioeconomy'. *Public Culture* 30 (3), pp. 413–439. https://doi.org/10.1215/08992363-6912115

Andrijasevic, R. and Mai, N. (2016) 'Editorial: Trafficking (in) representations: understanding the recurring appeal of victimhood and slavery in neoliberal times'. *Anti-Trafficking Review* 7, pp. 1–10. https://doi.org/10.14197/atr.20121771

Appadurai, A. (1996) *Modernity at Large: Cultural Dimensions of Globalization*. Minneapolis: Minnesota University Press.

Aradau, C. (2004) 'The perverse politics of four-letter words: risk and pity in the securitisation of human trafficking'. *Millennium: Journal of International Studies* 33 (2), pp. 251–277. https://doi.org/10.1177/03058298040330020101

Arata, S. (1996) *Fictions of Loss in the Victorian Fin de Siècle: Identity and Empire*. Cambridge: Cambridge University Press.

Arjana, S. R. (2015) *Muslim in the Western Imagination*. Oxford: Oxford University Press. http://doi.org/10.1093/acprof:oso/9780199324927.001.0001

Armiero, M. and Tucker, R. (2017) *Environmental History of Modern Migrations*. London: Routledge.

Asad, T. (1973) *Anthropology and the Colonial Encounter*. London: Ithaca Press.

Asad, T. (2007) *On Suicide Bombing*. New York: Columbia University Press.

Baishya, A. K. (2011) 'Trauma, post-apocalyptic science fiction and the post-human'. *Wide Screen* 3 (1), pp. 1–25.

Balibar, E. (2002) *Politics and the Other Scene*. London: Verso.

Barad, K. (1995) 'Meeting the universe halfway. Ambiguities, discontinuities, quantum subjects, and multiple positioning in feminism and physics'. In: Hankinson Nelson, L. and Nelson, J. (eds) *Feminism, Science and the Philosophy of Science*. Norwell: Kluwer Press, pp. 161–194.

Barad, K. (2008) 'Posthumanist performativity: toward an understanding of how matter comes to matter'. In: Alaimo, S. and Hekman, S. (eds.) *Material Feminisms*. Indianapolis: Indiana University Press, pp. 120–136.

Bauman, Z. (1997) 'The making and unmaking of strangers'. In: Werbner, P. and Modood, T. (eds) *Debating Cultural Hybridity: Multi-Cultural Identities and the Politics of Anti-Racism*. London: Zed Books, pp. 46–57.

Benjamin, W. (1940) 'Theses on the philosophy of history'. In: Benjamin, W. *Illuminations: Essays and Reflections*. New York: Schocken Books, pp. 253–264 (this edition 1968).

Benton, L. (2010) *A Search for Sovereignty: Law and Geography in European Empires 1400–1900*. Cambridge, UK: Cambridge University Press. https://doi.org/10.1017/CBO9780511988905

Bettini, G. (2013) 'Climate barbarians at the gate? A critique of apocalyptic narratives on "climate refugees"'. *Geoforum* 45, pp. 63–72. https://doi.org/10.1016/j.geoforum.2012.09.009

Bettini, G. (2017) 'Unsettling futures: climate change, migration and the (ob)scene biopolitics of resilience'. In: Baldwin, A. and Bettini, G. (eds) *Life Adrift: Climate Change, Migration, Critique*. London and New York: Rowman & Littlefield International, pp. 79–96.

Boeyink, C. T. (2019) 'A 'worthy' refugee: cash as a diagnostic of "xeno-racism" and "bio-legitimacy"'. *Refuge: Canada's Journal on Refugees* 35 (1), pp. 61–71. https://doi.org/10.7202/1060675ar

Boulila, S. C. and Carri, C. (2017) 'On Cologne: gender, migration and unacknowledged racisms in Germany'. *European Journal of Women's Studies* 24 (3), pp. 286–293. https://doi.org/10.1177/1350506817712447

Braidotti, R. (1997) 'Mothers, monsters, and machines'. In: Conboy, K. and Medina, N. (eds) *Writing on the Body: Female Embodiment and Feminist Theory*. New York: Columbia University Press, pp. 59–79.

Breidenbach, J. and Zukrigl, I. (2000) *Tanz der Kulturen. Kulturelle Identität in einer globalisierten Welt*. Hamburg: Rowohlt Verlage.

Brown, W. (1995) *States of Injuries: Power and Freedom in Late Modernity*. Princeton: Princeton University Press.

Brown, W. (2010) *Walled States, Waning Sovereignty*. New York: Zome Books.

Brown, W. (2017) 'Climate change, democracy and crises of humanism'. In: Baldwin, A. and Bettini, G. (eds.) *Life Adrift: Climate Change, Migration, Critique*. London and New York: Rowman & Littlefield, pp. 25–40.

Burroughs, E. R. (1911) *Under the moon of Mars*. New York: Simon & Schuster Books (this edition 2013).

Butler, J. (1993) 'Engendered/endangered: schematic racism and white paranoia'. In: Gooding-Williams, R. (ed.) *Reading Rodney King, Reading Urban Uprising*. New York and London: Routledge, pp. 15–22.

Butler, J. (2009) *Frames of War. When Is Life Grievable?* New York: Verso.

Chamayou, G. (2010) *Manhunts. A Philosophical History*. Princeton, NJ: Princeton University Press (this edition 2012).

Chambers, I. (2008) *Mediterranean Crossings. The Politics of an Interrupted Modernity*. Durham, Duke University Press. https://doi.org/10.1215/9780822388869

Chouliaraki, L. and Stolic, T. (2017) 'Rethinking media responsibility in the refugee "crisis": a visual typology of European news'. *Media, Culture and Society* 39 (8), pp. 1162–1177. https://doi.org/10.1177/0163443717726163

Chouliaraki, L. and Zaborowski, R. (2017) 'Voice and community in the 2015 refugee crisis: A content analysis of news coverage in eight European countries'. *The International Communication Gazette* 79 (6–7), pp. 613–635. https://doi.org/10.1177/1748048517727173

Clifford, J. (1986) 'Partial truths'. In: Clifford, J. and Marcus, G. E. (eds) *Writing Culture: The Poetics and Politics of Ethnography*. Cambridge, UK: Cambridge University Press.

Clifford, J. (1988) *The Predicament of Culture: Twentieth-Century Ethnography, Literature, and Art*. Cambridge, MA: Harvard University Press.

Cole, A. M. (2006) *The Cult of True Victimhood: from the War on Welfare to the War on Terror*. Stanford: Stanford University Press.

Cole, P. (2000) *Philosophies of Exclusion. Liberal Political Theory and Immigration*. Edinburgh: Edinburgh University Press.

Cranston, S., Schapendonk, J. and Spaan, E. (eds) (2018) 'New directions in exploring the migration industries'. *Journal of Ethnic and Migration Studies* Special Issue 44 (4), pp. 543–695. https://doi.org/10.1080/1369183X.2017.1315504

Creed, B. (1993) *The Monstrous-Feminine: Film, Feminism, Psychoanalysis*. London: Routledge. https://doi.org/10.4324/9780203820513

Creed, B. (2000) 'Alien and the monstrous-feminine'. In: Kirkup, G., Hovenden, F., Janes, L. and Woodward, K. (eds) *The Gendered Cyborg: A Reader*. London and New York: Routledge, pp. 110–135.

Danewid, I. (2017) 'White innocence in the Black Mediterranean: hospitality and the erasure of history'. *Third World Quarterly* 38 (7), pp. 1674–1689. https://doi.org/10.1080/01436597.2017.1331123

De Certeau, M. (1975) *The Writing of History*. New York: Columbia University Press (this edition 1988).

De Genova, N. (2013) 'Spectacles of migrant 'illegality': the scene of exclusion, the obscene of inclusion'. *Ethnic and Racial Studies* 36 (7), pp. 1180–1198. https://doi.org/10.1080/01419870.2013.783710

Del Lucchese, F. (2019) *Monstrosity and Philosophy: Radical Otherness in Greek and Latin Culture*. Edinburgh: Edinburgh University Press.

Dietze, G. (2016) 'Ethnosexismus. Sex-mob-narrative um die Kölner Sylvesternacht'. *movements. Journal for Critical Migration and Border Regime Studies* 2 (1), pp. 177–185.

Dixon, T., Hawkins, S., Heijbroek, L., Juan-Torres, M. and Demoures, F. X. (2018) *Attitudes towards National Identity, Immigration and Refugees in Italy*. [online] More in Common.

Dussel, E. (1995) *The Invention of the Americas. Eclipse of 'the Other' and the Myth of Modernity*. New York: Continuum.

Faber, M. (2000) *Under the Skin*. Edinburgh: Canongate Books.

Fabian, J. (1992) *Time and the Work of Anthropology: Critical Essays 1971–1991*. Switzerland: Harwood Academic Publishers.

Fallaci, O. (2006) *The Force of Reason*. Milan: Rizzoli.

Fanon, F. (1961) *The Wretched of the Earth*. New York: Grove Press (this edition 2005).

Farinelli, F. (2009) *La crisi della ragione cartografica*. Turin: Einaudi.

Farris, S. R. (2017) *In the Name of Women's Rights: The Rise of Femonationalism*. Durham: Duke University Press. https://doi.org/10.1215/9780822372929

Fassin, D. (2011) *Humanitarian Reason*. Berkeley, Los Angeles and London: University of California Press.

Federici, S. (2004) *Caliban and the Witch: Women, the Body and Primitive Accumulation*. Brooklyn, NY: Autonomedia.

Fekete, L. (2001) 'The emergence of xeno-racism'. *Race & Class* 43 (2), pp. 23–40. https://doi.org/10.1177/0306396801432003

Fekete, L. (2009) *A Suitable Enemy. Racism, Migration and Islamophobia in Europe*. New York: Pluto Press.

Finney, J. (1955) *The Body Snatchers*. New York: Dell Publishing.

Flammerion, N. C. (1862) *La Pluralité des mondes habités*. Paris: Hachette (this edition 2012).

Flammerion, N. C. (1864) *Les Mondes imaginaires et les mondes réels*. Paris: Hachette (this edition 2018).

Foucault, M. (1961) *History of Madness*. London and New York: Routledge (this edition 2006).

Foucault, M. (1975) *Discipline and Punish: The Birth of the Prison*. New York: Penguin (this edition 1991).

Foucault, M. (1976) *The History of Sexuality: 1: The Will to Knowledge*. New York: Penguin (this edition 1998).

Foucault, M. (1999) *Abnormal: Lectures at the College de France 1974–1975*. London and New York: Verso (this edition 2003).

Frassinelli, P. P. (2013) 'Il nuovo Sudafrica e gli alieni'. *Studi culturali* 2 (2), pp. 401–418.

Friedman, J. B. (1981) *The Monstrous Races in Medieval Art and Thought*. Cambridge, MA: and London: Harvard University Press.

Ghandi, L. (1998) *Postcolonial Theory A Critical Introduction*. New York: Columbia University Press.

Giametta, C. (2017) *The Sexual Politics of Asylum*. London: Routledge.

Giubilaro, C. (2017) '(Un)framing Lampedusa. Regimes of visibility and the politics of affect in Italian media representations'. In: Proglio, G. and Odasso, L. (eds) *Border Lampedusa. Subjectivity, Visibility and Memory in Stories of Sea and Land*. Cham: Palgrave Macmillan, pp. 103–117. http://doi.org/10.1007/978-3-319-59330-2_5

Giubilaro, C. (2018) 'Lo spettacolo del naufragio. Migrazioni, luoghi visuali e politica delle emozioni'. In: Arfini, E. A. G., Deplano, V., Frisina, A., Giuliani, G., Tesfaù, M. G., Perilli, V. et al. (eds) *Visualità & (anti)razzismo*. Padova: Padova University Press, pp. 10–23.

Giuliani, G. (2015) 'Fears of disaster and (post-)human raciologies in European popular culture (2001–2013)'. *Culture Unbound* 7 (3), pp. 363–385. http://doi.org/10.3384/cu.2000.1525.1572363

Giuliani, G. (2016a) *Zombi, alieni e mutanti. Le paure dall'11 settembre ad oggi*. Florence-Milan: Le Monnier /Mondadori Education.

Giuliani, G. (2016b) 'Monstrosity, abjection, and Europe in the War on Terror'. *Capitalism Nature Socialism* 27 (4), pp. 96–114. http://doi.org/10.1080/10455752.2016.1192212

Giuliani, G. (2017a) 'The colour(s) of Lampedusa'. In: Proglio, G. and Odasso, L. (eds) *Border Lampedusa*. Cham: Palgrave Macmillan, pp. 67–85. http://doi.org/10.1007/978-3-319-59330-2_5

Giuliani, G. (2017b) 'Afterword: life adrift in a postcolonial world'. In: Baldwin, A. and Bettini, G. (eds) *Life Adrift: Climate Change, Migration, Critique*. London and New York: Rowman & Littlefield International, pp. 227–242.

Giuliani, G. (2019) *Race, Nation and Gender in Modern Italy. Intersectional Representations in Visual Culture*. London: Palgrave Macmillan UK. http://doi.org/10.1057/978-1-137-50917-8

Giuliani, G. (2020 forthcoming) 'The end of the world as we know it'. In: Giuliani, G. (ed.) 'Cidadania em perigo: crime, fim-do-mundo e biopolítica nas literaturas e no cinema pós-coloniais'. *ecadernos* 32.

Giuliani, G., Giametta, C. and Petrovich Njegosh, T. (2020a forthcoming) 'Per un'analisi della memoria delle migrazioni in Europa: discorsi, (auto)rappresentazioni e propaganda'. In: Violi, P. and Salerno, D. (eds) *Migranti, archivi, patrimonio. Memorie pubbliche delle migrazioni*. Bologna: il Mulino.

Giuliani, G., Garraio, J. and Santos, S. J. (2020b) 'Online social media and the construction of sexual moral panic around migrants in Europe'. *Socioscapes* 1, pp. 161–180.

Graham, E. L. (2002) *Representations of the Post/Human. Monsters, Aliens and Others in Popular Culture*. New Brunswick: Rutgers.

Graham, S. (2011) *Cities Under Siege: the New Military Urbanism*. London: Verso.

Gregory, D. (1994) *Geographical Imaginations*. Cambridge, MA, and Oxford, UK: Blackwell.

Gregory, D. and Pred, A. (2006) *Violent Geographies: Fear, Terror, and Political Violence*. Abingdon and New York: Routledge. https://doi.org/10.4324/9780203944585

Guillaumin, C. (1972) *L'idéologie raciste. Genèse et langage actuel*. Paris: Mouton & Co.
Hage, G. (1998) *White Nation: Fantasies of White Supremacy in a Multicultural Society*. Annandale, NSW: Pluto Press.
Hage, G. (2003) *Against Paranoid Nationalism. Searching for Hope in a Shrinking Society*. Annandale, NSW: Pluto Press.
Hage, G. (2016) 'Etat de siège: A dying domesticating colonialism?' *American Ethnologist* 43 (1), pp. 38–49. http://doi.org/10.1111/amet.12261
Hall, S. (1996) 'Racist ideologies and the media'. In: Marris, P. and Thornham, S. (eds) *Media Studies. A Reader*. Edinburgh: Edinburgh University Press, pp. 160–168.
Haraway, D. J. (2003) *The Companion Species Manifesto: Dogs, People, and Significant Otherness*. Chicago: Prickly Paradigm Press.
Haraway, D. J. (2007) *When Species Meet*. Minneapolis: University of Minnesota Press.
Haraway, D. J. (2016) *Staying with the Trouble. Making Kin in the Chthulucene*. Durham: Duke University Press. https://doi.org/10.1215/9780822373780
Harley, B. J. and Woodward, D. (eds) (1987) *The History of Cartography*, vol. 1: *Cartography in Prehistoric, Ancient, and Medieval Europe and the Mediterranean*. Chicago: University of Chicago Press.
Hartman, S. V. (1997) *Scenes of Subjection: Terror, Slavery, and Self-Making in Nineteenth-Century America*. Oxford: Oxford University Press.
Herrmann, F. (2019) 'Unbemerkte Botschaften. Wie Populismus in die Leitmedien einfließt'. In: Müller, M. and Precht, J. (eds) *Narrative des Populismus*. Wiesbaden: Springer VS, pp. 147–162.
Hobson, J. (2005) *Venus in the Dark: Blackness and Beauty in Popular Culture*. London: Routledge. http://doi.org/10.4324/9780203700563
Huggan, G. (2008) 'Decolonizing the map: postcolonialism, poststructuralism and the cartographic connection'. In: *Interdisciplinary Measures: Literature and the Future of Postcolonial Studies*. Liverpool: Liverpool University Press, pp. 21–33.
Irigaray, L. (1977) *This Sex Which Is Not One*. Ithaca: Cornell (this edition 2008).
Iveković, R. (2019) *Politiques de la traduction. Exercices de partage*. Marseille: TERRA-HN éditions Alterego.
Jones, R. (2016) *Violent Borders: Refugees and the Right to Move*. [ebook reader] London and New York: Verso.
King, S. (1985) *The Mist*. New York: Viking Press.
Kristeva, J. (1980) *Pouvoirs de l'horreur: Essai sur l'abjection*. Paris: Édition du Seuil.
Latour, B. (2018) *Down to Earth Politics in the New Climatic Regime*. [ebook reader] Cambridge, UK and Medford, MA: Polity Press.
Lemberg-Pedersen, M. (2018) 'Security, industry and migration in European border control'. In: Weinar, A., Bonjour, S. and Zhyznomirska, L. (eds) *The Routledge Handbook of the Politics of Migration in Europe*. London: Routledge.
Linebaugh, P. (1991) *The London Hanged: Crime and Civil Society in the Eighteenth Century*. London: Allen Lane.
Linebaugh, P. and Rediker, M. (2000) *The Many-Headed Hydra: Sailors, Slaves, Commoners, and the Hidden History of the Revolutionary Atlantic*. Boston: Beacon Press.
Lochak, D. (2007) *Face aux migrants: état de droit ou état de siège?* Paris: Editions Textuel.
Loomba, A. (1998) *Colonialism-postcolonialism*. London and New York: Routledge.
Lytton, E. B. (1871) *The Coming Race*. Hollow Earth: BiblioLife (this edition 2008).
Mai, N. (2014) 'Between embodied cosmopolitism and sexual humanitarianism. The fractal mobilities and subjectivities of migrants working in the sex industry'.

In: Anteby-Yemini, L., Baby-Collin, V. and Mazzella, S. (eds) *Borders, Mobilities and Migrations. Perspectives from the Mediterranean, 19th–21st Century*. Brussels, Bern, Berlin, Frankfurt am Main, New York, Oxford and Wien: Peter Lang, pp. 175–192. https://doi.org/10.3726/978-3-0352-6416-6

Mamdani, M. (1996) *Citizens and Subjects*. Princeton; Princeton University Press.

Marx, K. (1867) *Capital: A Critique of Political Economy*, Volume 1. London: Penguin Books (this edition 1990).

Mbembe, A. (2003) 'Necropolitics'. *Public Culture* 15 (1), pp. 11–40. https://doi.org/10.1215/08992363-15-1-11

McClintock, A. (1995) *Imperial Leather: Race, Gender and Sexuality in the Imperial Context*. London and New York: Routledge. https://doi.org/10.4324/9780203699546

McGrane, B. (1989) *Beyond Anthropology: Society and the Other*. New York: Columbia University Press. https://doi.org/10.7312/mcgr90482

Mezzadra, S. (2001) *Diritto di fuga. migrazioni, cittadinanza, globalizzazione*. Verona: Ombre corte.

Mezzadra, S. (2004) 'The right to escape'. *ephemera* 4 (3), pp. 267–285.

Mezzadra, S. and Neilson, B. (2013) *Border as Method, or, the Multiplication of Labor*. Durham, NC: Duke University Press.

Mezzadra, S. and Sakai, N. (2014) 'Introduction'. Special issue 'Politics', *Translation. A transdisciplinary Journal* 4, pp. 9–29.

Mirzoeff, D. (2011) *The Right to Look. A Counterhistory of Visuality*. Durham: Duke University Press.

Mittman, A. S. and Dendle, P. (2012) *The Ashgate Research Companion to Monsters and the Monstrous*. Farnham, UK and Burlington, US: Ashgate. https://doi.org/10.4324/9781315241197

Mittman, A. S. and Hensel, M. (2018a) *Classic Readings on Monster Theory: Demonstrare*. Amsterdam: ARC Humanities Press. http://doi.org/10.5040/9781641899482

Mittman, A. S. and Hensel, M. (2018b) *Primary Sources on Monsters: Demonstrare*. Amsterdam: ARC Humanities Press.

Moulier-Boutang, Y. (1998) *De l'esclavage au salariat; Économie historique du salariat bridé*. Paris: PUF.

Neocleous, M. (2005) *The Monstrous and the Dead: Burke, Marx, Fascism*. Chicago: University of Chicago Press.

Nuzzo, L. (2013) 'Foucault and the enigma of the monster'. *International Journal of Semiotic Law* 26 (1), pp. 55–72. https://doi.org/10.1007/s11196-012-9275-8

Nuzzo, L. (2018) *Il mostro di Foucault. Limite, legge, eccedenza*. Milan: Meltemi.

Nuzzo, L. (2019) 'Foucault, the monstrous and monstrosity'. In: Chiesi, G. M. and Spiegel, F. (eds) *Classical Literature and Posthumanism*. London and Oxford: Bloomsbury, pp. 31–40.

Olson, G. (2013) *Criminals as animals from Shakespeare to Lombroso*. Berlin and Boston: Walter de Gruyter & Co. https://doi.org/10.1515/9783110339840

Olsson, G. (2007) *Abysmal: A Critique of Cartographic Reason*. Chicago: University of Chicago Press.

Perera, S. (2009) *Australia and the Insular Imagination: Beaches, Borders, Boats, and Bodies*. New York: Palgrave. http://doi.org/10.1057/9780230103122

Pesole, E. (2015) 'Genere, "razza" e crisi albanese a Telenorba'. In Giuliani, G. (ed.) *Il colore della nazione*. Florence–Milan: Le Monnier/Mondadori Education.

Pogliano, A. (2019) *Media e immigrazione*. Rome: Carocci.

Portinaro, P. P. (1992) 'Materiali per una storicizzazione della coppia amico-nemico'. In: Morani, M., Portinaro, P. P. and Vitale, A. (eds) *Amicus (inimicus) hostis*. Milan: Giuffrè, 1992, pp. 219–310.

Povinelli, E. (2016) *Geontologies. A Requiem to Late Liberalism*. [ebook reader] Durham and London: Duke University Press. https://doi.org/10.1215/9780822373810

Puar, J. K. (2005) 'Queer times, queer assemblages'. *Social Text* 23 (3–4), pp. 121–140. https://doi.org/10.1215/01642472-23-3-4_84-85-121

Puar, J. K. and Rai, A. (2002) 'Monster, terrorist, fag: the War on Terrorism and the production of docile patriots'. *Social Text* 20 (3), pp. 117–148. https://doi.org/10.1215/01642472-20-3_72-117

Puwar, N. (2004) *Space Invaders: Race, Gender and Bodies Out of Place*. New York: Berg.

Rediker, M. (2007) *The Slave Ship: A Human History*. New York: Penguin.

Said, E. (1978) *Orientalism*. London: Routledge and Kegan Paul.

Salgari, E. (1907) *Le meraviglie del 2000*. Massa Carrara: Transeuropa (this edition 2011).

Salerno, D. (2016) 'Risky subjects in time of terror. A semiotic perspective on the security discourse in Europe'. *Versus* 123 (2), pp. 363–384. http://doi.org/10.14649/85508

Salerno, D. (2017) 'The politics of response to terror: the reshaping of community and immunity in the aftermath of 7 July 2005 London bombings'. *Social Semiotics* 27 (1), pp. 81–106. https://doi.org/10.1080/10350330.2016.1151681

Santos, C. (2016) *Unbecoming Female Monsters: Witches, Vampires and Virgins*. Lanham: KY, Lexington Press.

Santos, R., Roque, S. and Santos, S. J. (2018) 'De-securitising "the South in the North"? Gendered narratives on the refugee flows in the European mediascape'. *Contexto Internacional* 40 (3), pp. 453–477. https://doi.org/10.1590/s0102-8529.2018400300003

Santos, R., Garraio, J., Giuliani, G., Roque, S. and Santos, S. J. (2019) 'A "crise dos refugiados" nos media europeus: alteridades, securitização e desconstrução das alteridades'. *Mundo Crítico. Revista de Desenvolvimento e Cooperação* 4, pp. 100–110.

Sartori, G. (2000) *Pluralismo, multiculturalismo e estranei. Saggio sulla società multietnica*. Milan: Rizzoli.

Sartori, G. (2015) *La corsa verso il nulla. Dieci lezioni sulla nostra società in pericolo*. Milan: Mondadori.

Sassen, S. (1999) *Guests and Aliens*. New York: Polity Press.

Saucier, P. K. and Woods, T. P. (2014) 'Ex aqua: the Mediterranean basin, Africans on the move and the politics of policing'. *Theoria: The Journal of Social and Political Theory* 141 (December), pp. 55–75. https://doi.org/10.3167/th.2014.6114104

Sayyid, S. and Vakil, A. K. (2010) *Thinking through Islamophobia: Global Perspectives*. London: Hurst & Company.

Schmigall, G. (1981) *Shakespeare and the Courtly Aesthetic*. Berkeley: University of California Press.

Schmitt, C. (1932) *The Concept of the Political*. Chicago: University of Chicago Press (this edition 1996).

Schmitt, C. (1950) *The Nomos of the Earth in the International Law of Jus Publicum Europaeum*. New York: Telos Press (this edition 2003).

Scott, J. C. (1998) *Seeing Like a State: How Certain Schemes to Improve the Condition Have Failed*. New Haven, CT: Yale University Press.

Shakespeare, W. (1610–1611) *The Tempest*. Ware: Wordsworth Editions (this edition 2013).

Sharma, S. and Sharma, A. (2003) 'White paranoia: orientalism in the age of empire'. *Fashion Theory* 7 (4), pp. 301–318. https://doi.org/10.2752/136270403778051952

Sharpe, C. (2009) *Monstrous Intimacies: Making the Post-Slavery Subject*. Durham: Duke University Press. https://doi.org/10.1215/9780822391524

Shildrick, M. (2002) *Embodying the Monster: Encounters with the Vulnerable Self*. London: Sage. http://dx.doi.org/10.4135/9781446220573

Simmel, G. (1903; 1908) 'Space and the spatial ordering of society'. In: Simmel, G. *Sociology. Inquiries into the Construction of Social Forms*. Volume 1. Leiden and Boston: Brill, pp. 543–621 (this edition 2009).

Sivanandan, A. (2001) 'Poverty is the new black'. *Race & Class* 43 (2), pp. 1–5. https://doi.org/10.1177/0306396801432001

Sontag, S. (1978) *Illness as a Metaphore*. New York: Farrar, Straus and Giroux.

Sousa Santos, B. (2014) *Epistemologies of the South. Justice against Epistemicide*. London: Paradigm. https://doi.org/10.4324/9781315634876

Sousa Santos, B. (2018) *The End of the Cognitive Empire. The Coming of Age of Epistemologies of the South*. Durham: Duke University Press. https://doi.org/10.1215/9781478002000

Sparke, M. (1998) 'Mapped bodies and disembodied maps: (dis) placing cartographic struggle in colonial Canada' In: Nast, H. and Pile, S. (eds) *Places Through the Body*. London: Routledge.

Spivak, G. C. (1988) 'Can the subaltern speak?' In: Nelson, C. and Grossberg, L. (eds) *Marxism and the Interpretation of Culture*. Houndmills: Macmillan, pp. 271–353.

Stolcke, V. (1995) 'Talking cultures: new boundaries, new rhetorics of exclusion in Europe'. *Current Anthopology* 36 (1), pp. 1–24. https://doi.org/10.1086/204339

Stoler, A. L. (2002) *Carnal Knowledge and Imperial Power: Race and the Intimate in Colonial Rule*. Berkeley: University of California Press.

Stoler, A. L. (2016) *Duress. Imperial Durabilities in Our Times*. Durham: Duke University Press. https://doi.org/10.1215/9780822373612

Svampa, M. (2019) 'El Antropoceno como diagnóstico y paradigma. Lecturas globales desde el Sur'. *Estudios. Utopía e praxis latinoamericana* 24 (84), pp. 33–54.

Tagg, J. (1988) *The Burden of Representation: Essays on Photographies and Histories*. Amherst: University of Massachusetts Press.

Tazzioli, M. (2017) 'Containment through mobility: migrants' spatial disobediences and the reshaping of control through the hotspot system'. *Journal of Ethnic and Migration Studies* 44 (16), pp. 2764–2779. https://doi.org/10.1080/1369183X.2017.1401514

Tazzioli, M. and Garelli, G. (2018) 'Containment beyond detention: the hotspot system and disrupted migration movements across Europe'. *Environment and Planning D: Society and Space*. https://doi.org/10.1177/0263775818759335

Torkington, K. and Perdigão Ribeiro, F. (2019) '*What are these people: migrants, immigrants, refugees?* Migration-related terminology and representations in Portuguese digital press headlines'. *Discourse, Context & Media* 27, pp. 22–31. https://doi.org/10.1016/j.dcm.2018.03.002

Trouillot, M. R. (1991) 'Anthropology and the savage slot: the poetics and politics of Otherness'. In: Fox, R. G. *Recapturing Anthropology: Working in the Present*. New Mexico: School of American Research Press, pp. 17–44.

Tsing, A. L. (2015) *The Mushroom at the End of the World: On the Possibility of Life in Capitalist Ruins*. Princeton, NJ: Princeton University Press.

Tuana, N. (2007) 'Viscous porosity. Witnessing Katrina'. In: Alaimo, S. and Hekman, S. (eds) *Material Feminisms*. Indianapolis: Indiana University Press, pp. 188–213.

Vila, P. (ed.) (2003) *Ethnography at the Border*. Minneapolis and London: University of Minnesota Press.

Weber, B. (2016) ' "We Must Talk About Cologne": Race, Gender, and Reconfigurations of 'Europe'. *German Politics and Society* 34 (4), pp. 68–86. https://doi.org/10.3167/gps.2016.340405

Wells, H. G. (1898) *War of the Worlds*. Manor, TX: Arc Manor (this edition 2008).

Wells, H. G. (1901) *The First Man on the Moon*. London: Penguin (this edition 2015).

Werthschulte, C. (2017) 'Nach' Köln ist wie "vor" Köln. Die Silvesternacht und ihre Folgen'. In: Bundeszentrale für politische Bildung. (ed.) *Köln. Aus Politik und Zeitgeschichte*, pp. 10–17. Available at: www.bpb.de/apuz/239696/die-silvesternacht-und-ihre-folgen. Accessed 5 August 2019.

Wolfe, C. (2005) *Monsters and Philosophy*. London: College Publications.

Yancy, G. (2012) *Look, A White! Philosophical Essays on Whiteness*. Philadelphia: Temple University Press.

Yusoff, K. (2018) *A Billion Black Anthropocenes or None*. [ebook reader] Minneapolis: Minnesota University Press.

Žižek, S. (2018) 'Are the worst really full of passionate intensity?'. *Statesman*. Available at: www.newstatesman.com/world-affairs/2015/01/slavoj-i-ek-charlie-hebdo-massacre-are-worst-really-full-passionate-intensity. Accessed 18 December 2019.

Žižek, S. (2016) *Against the Double Blackmail: Refugees, Terror and Other Troubles with the Neighbours*. New York: Penguin.

Films

Arnold, Jack. *It Came from Outer Space*. USA, 1953.

Blomkamp, Neil. *District 9*. South Africa, 2009.

Carpenter, John. *The Thing*. USA, 1982.

Carpenter, John. *They Live*. USA, 1988.

Dante, Joe. *Gremlins*. USA, 1984.

Darabont, Frank. *The Mist*. USA, 2007.

de Chomón, Segundo. *Excursion dans la Lune*. Spain, 1908.

del Toro, Guillermo. *The Shape of Water*. USA, 2017.

Edwards, Gareth. *Monsters*. UK, 2010.

Emmerich, Roland. *Independence Day*. USA, 1996.

Fowler, Gene Jr. *I Married a Monster from Outer Space*. USA, 1958.

Garland, Alex. *Annihilation*. USA, 2018.

Glazer, Jonathan. *Under the Skin*. UK, USA & Switzerland, 2013.

Hall, Stuart. *Race. The Floating Signifier*. UK, 1997.

Haskin, Byron. *The War of the Worlds*. USA, 1953.

Herek, Stephen R. *Critters*. USA, 1986.

Méliès, Georges. *Le Voyage dans la Lune*. France, 1902.

Méliès, Georges. *Voyage à travers l'impossible*. France, 1904.
Menzies, William C. *Invaders from Mars*. USA, 1953.
McCarthy, Colm. *The Girl with All the Gifts*. UK, 2016.
McTiernan, John. *Predator*. USA, 1987.
Neumann, Kurt. *Kronos*. USA, 1957.
Novelli, Enrico. *Un matrimonio interplanetario*. Italy, 1910.
Nyby, Christian. *The Thing from Another World*. USA, 1951.
Rilla, Wolf. *Village of the Damned*. USA, 1960.
Romero, George. *Dawn of the Dead*. USA, 1978.
Romero, George. *Land of the Dead*. USA, 2006.
Scott, Ridley. *Alien*. USA, 1979.
Spielberg, Steven. *Close Encounters of the Third Kind*. USA, 1977.
Spielberg, Steven. *E.T.* USA, 1982.
Spielberg, Steven. *War of the Worlds*. USA, 2005.
Sears, Fred F. *Earth vs. the Flying Saucers*. UK, 1956.
Siegel, Don. *Invasion of the Body Snatchers*. USA, 1956.
Stow, Percy. *When the Man in the Moon Seeks a Wife*. UK, 1908.
Villeneuve, Denis. *Arrival*. USA, 2016.
Wise, Robert. *The Day the Earth Stood Still*. USA, 1951.
Wyatt, Rupert. *Captive State*. USA, 2019.

TV series

AAVV. *Star Trek*, USA, 1966–1969; 1987–2005; 2017–.
Johnson, Kenneth. *Visitors*. USA, 1984–1985.
Torpe, Christian. *The Mist*. USA, 2017 (1 season, 10 episodes).

Art

Dumas, Magdalène. *The Image as Burden*. South Africa, 1995.
Cox, Renée Valerie and Lyle Ashton Harris. *Venus Hottentot 2000*. USA, 1995.
Ligon, Glenn. *Runaways*. USA, 1993.
Walker, Kara. *Endless Conundrum, an African American Anonymous Adventuress*, USA, 2001.
Walker, Kara. *A Subtlety, or the Marvelous Sugar Baby*, USA, 2014.

3 Lifting the veil on the monstrous Anthropocene

A postcolonial analysis

1 Introduction

Catastrophe is the moment of *revelation*, when the veil is lifted on the *monstrosity* of the Anthropocene. If we read this section in parallel with Sections 1 and 2, we would see how the monstrification of human mobility and terrorism in the safe space of the 'we' conceals the *monstrosity* of the Anthropocene, dissociating the environmental apocalypse from the history of horrors that accompanies it. A closer look at the effects of global inequality reveals instead how human exploitation, the objectification of non-human species, natural resource depletion, and land and water pollution originate in the same Anthropocenic ontologies and capitalist logic. At the same time, the unfolding catastrophe of the Anthropocene *reveals* the temporariness and precarity of those violent discursive *dispositifs* that in the previous sections I called *history* and *geography* – deployed since Modernity to safeguard the privileges of the 'we'. Fearing an environmental apocalypse is not about fearing Earth's extinction, because life on the planet will continue, as it always has:

> The rock record shows that after each mass extinction, the organismically interweaving biosphere has regrown to form more species, cell types, metabolic skills, areas settled, networked intelligences, and complex sensory skills than before. Maybe this time, instead of hurting it, we can help it continue its multispecies energy-transducing recycling ways for billions of years more.
>
> (Sagan 2017, p. 174)

It is instead about fearing the loss of those *dispositifs*, and that the dualisms of culture/nature, human/non-human, organic/inorganic, Here/Out there, civilised/barbarian, places *of* disaster/places *for* disaster, centre/peripheries of the world, together with their borders, walls, colonies and segregated spaces will not succeed in containing a catastrophe that has the potential to affect everyone. Modern Western definitions of humanness, worthiness, priorities, progress and autonomy as well as the concept of *the political* depend so heavily on such firmly established *dispositifs* (see Sections 1 and 2) that their disappearance is inconceivable, an omen of unrestrained barbarity and the eventual *end of humanity*. And yet, trapped in its own logic, when faced with its own vulnerability to the effects of the Anthropocene, the 'we' cannot but accelerate the catastrophe.

This section will focus on the *panic* triggered by the loss of *history* and *geography* on one hand and by the end of *humanity* on the other, tying together my discussion of fears of invasion and a backlash to Europe's past history of violence (Section 1) and my reflections on the restrictions to people's freedom of movement within 'racist' Capitalocene's great acceleration and multiplications of borders (Vergès 2017; see Section 2). I will reread my reflections in the frame of persisting fantasies of ultimate catastrophe and dualisms between barbarism and civilisation, reason and darkness, order and chaos underpinning the discourse of the war on terror and the de-politicisation of environmental issues. As in the previous section, I will base my analysis on the distinction between catastrophe and disaster. A catastrophe is a productive crisis – and, in line with Christian eschatology, is taken both as an omen of *the end of the world as the 'we' knows it* and as a promise of a new and better world. Disaster is perceived as uncontrollable when related to the Out there and its helpless peoples; it is instead a non-systemic and hence manageable event when occurring in the space of the 'we'. The same distinction applies to the dualism between places *of* disaster and places *for* disaster. The former, where disaster is a foreseeable and avoidable *occurrence*, are inhabited by the 'we'. The latter, where the 'we' believes disaster to be endemic and essentially linked to backwardness, poverty and lack of knowledge and skills, are inhabited by the non-civilised.

Public perception of the planet and the impending catastrophe is shaped by stable and reassuring Anthropocenic binaries providing an imaginary safety net against the threat of chaos. Accordingly, post-apocalyptic scenarios in popular culture are still very much anthropocentric in nature. A world devoid of humans and no longer ruled by Western rationality and techno-science is simply inconceivable, as we will see in Chapters 5, 6 and 7 where I analyse popular disaster and doomsday films.

In unpacking the Anthropocenic post-apocalyptic imaginary, I will build my critique on Povinelli's (2016) notions of *geontology* and *geontopower*, Alaimo's (2010) *trans-corporeality* and Barad's (2007) *intra-activity*: Povinelli's *geontopower* describes 'a set of discourse, affects, and tactics used in late liberalism to maintain or shape the coming relationship of the distinction between Life and Nonlife'. *Geontopower* is grounded in *geontologies* that present this distinction as the supremacy of human Life over non-human Life and Nonlife (the inorganic). Bio- and necropower, along with their respective bio- and necropolitics, hinge on this very distinction. In Alaimo's *trans-corporeality*,

> the human is always intermeshed with the more-than-human world, underlines the extent to which the substance of the human is ultimately inseparable from 'the environment'. It makes it difficult to pose nature

> as mere background, as Val Plumwood [1993] would put it, for the exploits of the human since 'nature' is always as close as one's own skin – perhaps even closer. Indeed, thinking across bodies may catalyze the recognition that the environment, which is too often imagined as inert, empty space or as a resource for human use, is, in fact, a world of fleshy beings with their own needs, claims, and actions. By emphasizing the movement across bodies, trans-corporeality reveals the interchanges and interconnections between various bodily natures.
>
> (2010, p. 2)

Barad's *intra-activity* refers to the making of objects, subjects and knowledge through intra-action such that 'the object and the measuring agencies emerge from, rather than precede, the intra-action that produces them' (2007, p. 128). In this framework, 'agency is a matter of intra-acting; it is an enactment, not something that someone or something has. Agency is doing/being in its intra-activity' (2007, p. 235). I will also apply to my analysis her notion of intra-active agentic knowledge production, according to which:

> Practices of knowing and being are not isolable; they are mutually implicated. We don't obtain knowledge by standing outside the world; we know because we are of the world. We are part of the world in its differential becoming. The separation of epistemology from ontology is a reverberation of a metaphysics that assumes an inherent difference between human and nonhuman, subject and object, mind and body, matter and discourse.
>
> (2007, p. 185)

Finally, I will borrow Barad's concept of *intra-activity* to understand human subjectivity as emanating from of a system of mutual *trans-corporeal* interdependencies (Alaimo 2010) – pre-existing interacting phenomena and socio-cultural conditions summarised by the concept of 'environment'. I will employ the concepts of *geontopower*, *geontologies*, *trans-corporeality* and *intra-activity* throughout this section, namely in Chapters 5, 6 and 7, to investigate the relationship between Life and Nonlife (Povinelli 2016) and between human life and racialised less-human life (see also Yusoff 2018; Davis et al. 2019) in the Anthropocene. More importantly, I will argue for a *politically* and *responsibly* engaged *trans-corporeal* and *intra-active* conception of the planet, one that recognises the interdependency between organic beings and inorganic formations, between the Here and the Out there, and between past, present and future (Barad 2008).

* * *

From colliding meteors and planets to viral pandemics and nuclear or biochemical wars, fantasies of extinction have long haunted humankind. And yet the Anthropocenic Anthropos has not slowed his pace, in the belief that *planet* Earth is isomorphic to the *world* of humans – as if Earth may only exist through the human gaze, thought, and action. Seen for centuries as divine punishment, catastrophe has been interpreted since the eighteenth century either as a symbol of the impossibility of mastering nature or, on the contrary, as nature's rebellion against human hubris – and an inherent risk in human technology and the 'acceleration' of progress in accord with the dualistic Cartesian schism between man and the environment. There has long been the idea that taking a step back or a leap forward along the timeline of civilisation is the only way for humankind to survive and even benefit from a crisis, either going back to the primitive life of the 'good savage' and giving civilisation a new start (Rousseau) or accelerating techno-scientific progress (Voltaire). Far from effectively challenging the dualism between technology and nature, these approaches entail a temporary escape to the past and limited revision of European civilisation models or require more sophisticated technology to better dominate nature.

From Jean Jacques Rousseau's 'good savage' to Terrence Malick's *The Thin Red Line* (1998), fantasies of returning pre-capitalist or colonial histories of violent progress have long served imaginaries of salvation and temporary escape from a world governed by anthropocentric rules. Inspired by these fantasies, the search for new beginnings away from religious wars and authoritarian monarchs prompted many a colonising enterprise but never truly challenged the imperatives of the Anthropocene, representing instead a mere pause to prepare for the next step in the value-extraction process, be it colonialism or slavery. Fantasies of escape often took the form of radical critiques of capitalism and state discipline, such as that embodied by mutineers and fugitive white indentured servants who would embrace alternative cosmogonies and kinships practised by indigenous peoples, wandering witches, or the enslaved (Snodgrass 2008; Yusoff 2018; Linebaugh and Rediker 2000). *The Thin Red Line* narrates one such story of mutiny with a pessimistic view of the impact of capitalism and Anthropocenic war mentality. Like many others in the past, the deserters in the film have left their world behind them to join the left-outs and their *pluriverse* (Escobar 2016; 2014) in one of Earth's last Edens, but there is no turning back from the apocalypse brought by the Second World War and colonialism. As for the 'good savage' in Western political thought and the Tahitians in the *Bounty* Mutiny (who like many others during colonial expansion were exoticised, converted to Christianity, raped and eventually exterminated),[1] rather than questioning colonial Anthropocene they clearly represent the enduring appropriation of the left-outs of Western 'civilisation'.

Unlike past imaginaries of catastrophe, present fantasies of planetary apocalypse involve tangible events that, for the first time in human history, are not spatially and temporally bounded but rather uncontrollable. Even Cold War fears of nuclear annihilation were mitigated by the awareness that humans could prevent the apocalypse if only they wanted.

But unlike nuclear bombs, the panoply of floods, tsunamis, toxic wastes and other disasters that have been affecting the life of the Other Out there since the 'Great acceleration' are seen now as *having no boundaries* and being capable of shattering the space of Western and white techno-scientific order, upsetting its social and cultural foundations. After a temporary phase in which camps, colonies, walls and carceral archipelagos were able to contain the 'flood' from the Out there, there are now widespread fears that an environmental apocalypse would eventually erase all differences, lowering all to the same common level and vulnerability to death – that is, wiping out all distinctions legitimising notions of *humanness*, *belonging*, and *identity* in modern times, as well as local and global *privileges*, *hierarchies* and *inequality*. As I argued in Section 2, the unbounded nature of environmental 'globalisation' generates fantasies of invasion and terror that sometimes turn into apocalyptic fears of a 'great replacement' and calls for tighter borders, military violence against the 'enemy', thicker collective (racialised) identities, and higher inequality to preserve the 'we' (Latour 2018). Fears need to be contained to maintain local and global order for the time being; however, a culture of fear is instrumental to perpetuating securitarian discourses and the need for a security apparatus to provide reassurance that the 'we' and the Here are still physically distanced from and protected against the side effects of the Anthropocene.

Moreover, there are fears that a planetary catastrophe would instantly and *irrevocably* erase time (or *history*) and a new 'empty temporality' would frustrate all efforts at building for the future by breaking the linear perception of time that, regardless of the crisis of the idea of progress, is still ingrained in the anthropocentric understanding of a fully developed human nature. In fact, the environmental apocalypse will not erase time but rather spell the end of anthropocentric time – that same linear perception of time that has solidified the belief in Western techno-scientific superiority, relieving the 'we' of all responsibility for the horrors of the past (Section 1).

Finally, unlike fictional narratives such as *The Core* (2003), *Armageddon* (1998) and *The Wandering Earth* (2019) where the apocalypse is averted through man's intervention, whether by setting off nuclear devices to jump-start the Earth's rotation, bombing a meteor that threatens to destroy the planet, or moving Earth to a different planetary system altogether, there is the perception that the apocalypse is an *inexorable* event – however unacknowledged by those who could postpone the end of days. As in Cormac McCarthy's *The Road* (2006) – in which father and son try to survive a post-nuclear scenario – what is feared is not desertification, but that nature will not survive an apocalypse that will necessarily swallow the human 'world', together with what is believed to be the only reason for the planet to exist: human life.

To dispel these fears, sci-fi offers soothing fantasies of '(white) saviours of civilisation' able to either prevent or undo the damage caused by a catastrophe, or new planets and space habitats where capitalism can be rebuilt and with it *the world as the 'we' knew it*. Similarly, the mainstream discourse of climate change tends to underestimate the lasting effects of environmental catastrophe, focusing more on depoliticising and normalising turbo-capitalism and postcolonial divides than providing a deep critique of the Anthropocene narrative (Bettini 2013, p. 69; Swyngedouw and Ernstson 2018). In fact, what the 'we' wants is to be discharged of all guilt, as well as being reassured that survival is possible and pre-apocalyptic time and life can be rebuilt. The 'we' wants to be the reason why *Terminator* (1984, 1991, 2003, 2009, 2019) keeps coming back from the future to kill the child who will save humankind. The 'we' wants to be the father who in *Interstellar* reaches back in time from a parallel dimension to tell his daughter that there is still hope. In the film, Joseph Cooper (Matthew McConaughey), an engineer and former NASA pilot, reveals that the *monstrous Anthropocene* cannot be stopped: the only way to escape climate-change effects, contaminated crops and a planetary famine is to look for a second chance on a different planet. There is no room left for reflection on the 'we's' responsibilities in a narrative that presents knowledge and technology as the only means for *humanity* to start over and leave behind an environmental catastrophe caused by capitalism and value extraction. The Anthropocene will live on, although it is:

> clearly not the product of 'human nature,' or humanity as a whole, but rather interrelated historical processes set in motion by a small minority. This privileged cadre provided the preconditions for the development of global capitalism through processes of settler colonialism and enslavement, organized and rationalized by racism. Consequently, the centering of an undifferentiated humanity in much Anthropocene scholarship serves to reproduce white supremacist claims to universal knowledge. Moreover, it implies that the Anthropocene heralds a post-racial future of generalized climate catastrophe and the dissolution of binary categories delineating 'humanity' and 'nature' (see Leong 2016; Luke, 2018; Mirzoeff 2018).
>
> (Davis et al. 2019, p. 4)

Not only does human survival downgrade catastrophe to a crisis that can be overcome, but it makes the space for critique of the Anthropocene smaller. The result is a widespread perception that the Anthropocene is not as bad as some paint it to be and, in the end, only needs a few tweaks: all the 'we' needs to do is improve technology for a more humane exploitation, more efficient colonisation and more sustainable extraction – that is, for a more gentle destruction of Life and Nonlife. When faced with the possible end of the planet, the monster seems to demand yet more *monstrous intimacies* (Sharpe 2009): more dualisms, techno-science, capitalism, plantations, camps, walls, colonies, carceral archipelagos, borders and boundaries. Fantasies of post-apocalyptic *renewal* that are grounded in 'colonial' and 'cultural' (national) archives (Stoler 2016; Wekker 2016, respectively) and triggered by the crisis of capitalism signalled by the catastrophe (Saldanha 2017) can only rehash the same old capitalist logic, hierarchies and inequality as the Anthropocene. Not all sci-fi films reproduce these fantasies. Chapters 6 and 7 present mostly dystopian narratives (except for *The Impossible*) that highlight global inequality both in the present and in the post-apocalyptic future, while narratives of the '(white) saviour of civilisation' (Chapter 5, except for *Snowpiercer*) are apologetic of the global order as it is. Dystopian films are set in a post-apocalyptic world in which today's inequality is taken to the extreme: some humans will be left behind in a dying planet, others will undergo mutation to survive, and a few will be chosen to exploit new lands. While the masses of the wretched poor and postcolonial subjects are segregated, detained or deported (*Code 46*) and left to deal with famine, deadly sunrays and desertification, only a selected few are allowed to share in the comforts and safety of hyper-technological global cities. The Brazilian sci-fi drama *3%* (2016–) follows the same narrative path: upon turning 20 years old, healthy, intelligent, fertile and assertive slum residents are given a chance to leave squalor behind and join the elite in the 'other-world'.

This section is organised as follows. Chapter 2 focuses on the notion of catastrophe and its genealogy. Concepts of punishment and *renewal* are explored in the history of Western political thought in the wake of the shift from theological to secular ideas of nature and catastrophe and read contrapuntally through the indigenous idea of colonisation as the ultimate catastrophe. Chapter 3 re-thinks Anthropocenic dualisms of Here and Out there, civilisation and barbarity, places *of* disaster and places *for* disaster, worthy and expendable, meaningful and meaningless in the light of recent human and environmental disasters and the frantic quest for new havens, camps and colonies. Chapter 4 focuses on resilience, resistance and struggle in the context of both the mainstream discourse of environmental catastrophe and radical theories of degrowth and social reproduction, providing a critical analysis of the interdependency between care, self-care and earth-care. Self-care is understood here in terms of black feminist and queer critique and re-appropriation of the term as a necessary *political* tool for deconstructing the *patriarchal* model of care. Self-care, then, is not seen as a 'neoliberal trap' (Michaeli 2017) but rather understood within the frame of a *political* conception of the environment as *trans-corporeal* and *intra-active* – that is, based on mutual affection and interdependency among human and non-human beings and inorganic formations (Barad 2008). *Human responsibility* towards Life and Nonlife becomes then a political value. Chapters 5, 6 and 7 focus on post-apocalyptic films grouped according to a common scenario: the trope of the (white) saviour of civilisation; world *renewal* through human selection and enhancement; and, finally, a new world in which humans have undergone permanent metamorphosis. The first group of films includes *The Core* (2003) by Jon Amiel, *The Wandering Earth* (2019) by Frant Gwo, *The Day After Tomorrow* (2004) by Roland Emmerich, and *Snowpiercer* (2013) by Boong Joon Ho. The second scenario will be analysed through films and TV series such as *The Impossible* (2011) by Juan Antonio Bayona, *Blindness* (2008) by Fernando Mereilles, *Elysium* (2013) by Neill Blomkamp, *I Am Mother* (2019) by Grant Sputore, *Code 46* (2003) by Michael Winterbottom and the Web television series *3%* (2016–) by Pedro Aguilera. The third scenario will be analysed through the film *The Girl with All the Gifts* (2016) by Colm McCarthy. In Chapter 8, I will sum up my conclusions.

2 The drowned and the saved in the Anthropocene

The association between catastrophe and divine punishment is seemingly still popular. *The Mist* evokes the conflict between Christian fundamentalism and secularism in the face of an extraordinary catastrophe, presenting two contrasting views on the 'invasion', which some believe to be God's punishment while others ascribe to 'bad' progress. And although the secularist view is privileged, the film's vision of the future is not optimistic, because the same progress that allowed monsters to spill into the world through a door left open by the military has also meant the death of society's youngest and most vulnerable members, embodied in the film by Billy, the protagonist's son. Neither the biblical nor the secular vision of the apocalypse, however, seems to express an interest in the fate of non-human beings and inorganic formations. Only humans can be damned or saved, because they have a closer relation to God than any other creature or, in secular narratives, represent the more advanced link in the great chain of being.

Tracing the genealogy of the image of Apocalypse from early Christianity to the Enlightenment can help understand what meanings shape today's imaginary of (environmental) crisis. In Christian eschatology as well as in Jewish and Islamic holy scriptures, the Apocalypse has been described as the day of revelation and judgment. In Saint John's *Book of Revelation*, the opening of the seven seals marks the second coming of Christ and the beginning of the Apocalypse: the great catastrophic time of trouble culminates in the battle of Armageddon at the end of days that brings human rulership to an end and is followed by the creation of 'a new heaven and a new earth'. Eschatological interpretations abound in the TV series *In the Flesh* (see Section 1), in which the resurrection of the dead is presented both in biophysical and in religious terms. While scientists are busy examining and 'rehabilitating' those affected by PDS (partially deceased syndrome), the church-going residents of Roarton see it either as a sign of the impending return of Christ or as the ominous warning of the looming Apocalypse.

In general, it was not until the Renaissance and Enlightenment that natural catastrophes stopped being seen as God's punishment and a warning about the end of days. Since then, the word catastrophe has gained new meanings thanks to contributions from geology, astronomy, physics and biology. Beside scientists, philosophers also began basing their knowledge claims on 'the hypothesis of a rationalised, controlled (some would say domesticated) nature that is abruptly surrendered to the mastery of science and technology, which deplete its resources'[2] (Velescu 2017, p. 50). This hypothesis was shared by the *philosophes des Lumières*, although with different and often conflicting interpretations. Montesquieu 'favoured [...] the idea of a catastrophic destruction that would lead human society to save itself thanks to rational thinking and scientific progress'[3] (Velescu 2017, p. 60). Voltaire (1756; 1759) saw the absence of reason in nature as a manifestation of evil and Rousseau (1755), for whom society corrupted nature's work, regarded natural disasters, and the Lisbon earthquake in particular, as nature's warning signs about human wrongs and thus as a chance for *social renewal*. According to Diderot's (1770; 1780–1782) *déterminisme climatique*, the potential for a catastrophe to produce social change owed much to a people's national character. Condorcet (1795) thought that 'Nature and the physical laws of the universe might be responsible for impediments to progress'[4] (Velescu 2017, p. 63). However, as he argued, 'the progress of this perfectibility [is] above the control of every power that would impede it'[5] and a catastrophe always allows humans to 'be a part of history'[6] and set civilisation in motion (1795, p. 81; Mercier-Faivre and Thomas 2008).

Apparently, although

> a few [still] saw the earthquake as a triumph for the Jansenists, since the quake had crushed the center of Jesuit power [;] Protestants could see the quake as a lesson for Catholics, and both Protestants and Catholics could see the quake as directed towards wickedness and towards the inquisition.
>
> (Dynes 2000, p. 99)

the idea that catastrophe could be an engine of social 'improvement' was to a certain degree shared by European intellectuals. The panoply of urban projects on display at the 'Creation from Catastrophe' exhibition at The Architecture Gallery in London (2015)[7] seems to confirm that after each earthquake, dozens of European artists, intellectuals, architects and geographers have rushed to redesign urban space so as to achieve a more rational social organisation.

In general, as Mercier-Faivre and Thomas (2008) argue, European philosophers in the *siècle des Lumières* drew a distinction between natural disasters and political catastrophes, framing the former as a matter *beyond politics* and associating the latter with revolutions. In fact, in Western early modern political thought, political apocalypses were only associated with religious wars, revolutions and regicide, against which utopianism between the 1500s and the 1650s had imagined new islands, cities, and unknown

Figure 3.1 João Glama, 'O Terramoto de 1755' (fragment), oil on canvas, c. 1756–1792.

lands where to find refuge and build a more equal and pious society. That was the case with Thomas Moore's *Utopia* (1516), Ludovico Agostini's *Imaginary Republic* (1580), Tommaso Campanella's *The City of the Sun* (1623), Francis Bacon's *The New Atlantis* (1627), and finally James Harrington's *The Commonwealth of Oceana* (1656). Their legacy is to be found as much in early twentieth-century sci-fi fantasies of conquering new planets (Section 2) as in astrophysical experiments, whose common denominator was an actual or imaginary point zero where human history could begin anew. New Utopias would allow the white Anthropos to build an alternative to the Anthropocene or, on the contrary, to better meet those civilisational standards legitimating capitalism, (hetero)patriarchy and colonialism.

Owing to the belief that the environmental issue was 'beyond politics' and 'beyond religion', the planet was reduced to the object of human reason and a commodified source of value extraction. The environment's life was considered of secondary importance to human life and relegated to a subordinate position in *history* and *geography*. Ruling over nature became a moral imperative: referring to the Lisbon earthquake, Immanuel Kant stated that 'Nature is the exact opposite of freedom' (cited in Clark 2011, p. 90). At the same time, for the German philosopher as for Rousseau, humankind was in a necessary relation to nature. In order to survive, they argued, humans should conform their actions to nature, and not the contrary: 'Kant notes that it would be better to build smaller buildings and use lightweight building materials in earthquake-prone areas' (Velescu 2017, p. 69).[8] As in the case of the Lisbon earthquake that he discussed extensively in his early writings, catastrophic events should be regarded as a means for improving human technique and society (Velescu 2017; Tagliapietra 2004).

As Mercier-Faivre and Thomas (2008) argue referring to the philosophical debate on the plague epidemic in Marseille (1720) and the Lisbon (1755) and Messina (1783) earthquakes, this meant that catastrophes could provide the opportunity to improve social organisation and the bio- and necropolitics of the time. At the same time, the economy could be improved by more effective exploitation and extraction of organic and inorganic resources – that is, encouraging *human revolution over nature* and letting technique (and value extraction) rule the world. Only by doing so, Kant believed, would human autonomy and self-responsibility be accomplished.

And indeed, in his view the constraints nature places on human activity are contextual and the associated risks are predictable – sometimes those risks can be avoided, and at times they are even productive of new local and global orders. What is striking is that Kant's human ontology, always promising new utopian republics where a life not as fraught with *political* and *environmental pollution* is possible, is based on those same 'pathological tendencies of self-directing and world-mastering subjectivity' (Clark 2011, p. 100) that today are increasingly associated with the risk of ultimate catastrophe. In the eighteenth century, nature was not only a limit but also an endless resource. The Enlightenment belief in limitless human advancement was at the core of this conception, which did not include the possibility that progress could lead to an environmental catastrophe, although this was already a reality for those who were forced to pay the price for Western development. This view of nature persisted for over two centuries and went on to become the fundament of late modern techno-scientific research and politics. Even in the face of a planetary crisis that could lead to the extinction of the Anthropocenic Anthropos and innumerable other species, the dominant imaginary is the same today as it was in the past: *profit maximisation and the belief in linear progress will save us from the deadly effects of profit maximisation and linear progress.* Despite this being a crisis of modern reason and subjectivity whose urgency has been proven by science – the highest expression of Western civilisation, and 'pivotal in creating the field of visibility and the planetary glance necessary to think of environmental catastrophe' (Bettini 2020) – *more Anthropocene* is still the dominant response.

As the American historian Kevin Rozario (2007) shows, Western ideas of catastrophe as a source of moral, cultural and political *renewal* are still very popular. My point is that the modern anthropocentric and Anthropocenic view of catastrophe as a means to *renewal* is rearticulated today in public discourse and apocalyptic sci-fi to envision a reassuring, utopian, eugenic future, which also corresponds to a new 'beginning' in the context of the growing imaginary of *crisis*. As Laura Pulido (2018) argues, public discourse on climate change does not explicitly mention these *dispositifs* but rather silently advocates the expendability of the unworthy:

> At the close of the United Nations climate summit in Copenhagen in December 2009, governments agreed to a global temperature increase of two degree Celsius. It was thought that two degrees would prevent global catastrophe. However, it was fully understood that two degrees would eliminate some island states and be absolutely disastrous for much of Africa. This is key: *knowingly* allowing large swaths of nonwhite, mostly poor people to die. Could we have decided to do otherwise? Yes. But as a global community we have declined to prevent this massive die-off. In response, a group of African delegates expressed their outrage, protesting: 'We will not die quietly', 'Two degrees is suicide', '1.5 to Survive', and 'Death sentence for Africa'. The delegates were not about to go quietly and wanted to ensure that everyone was aware of the import of their actions. This moment illuminated the racial geography of global warming, our lack of political will, our disregard for nonwhite and poor lives, and the deeply immoral nature of the Anthropocene.
>
> (p. 120)

Second World War atrocities and Nazi eugenics still cast a long shadow over society, and practices of human selection and manipulation of those deemed superior, or the killing of those deemed less worthy, are a totally unacceptable topic. But post-apocalyptic sci-fi is not bound by the constraints of political correctness and therefore can explicitly suggest 'memories of the future' that establish connections between the post-apocalyptic future and a not-so-distant past of colonial and totalitarian bio- and necropolitics of killing or eugenic reproduction. Furthermore, the 'natural' elements of catastrophe preclude accountability, providing an expedient way to ensure the survival or advancement of the (human) species by preserving its superior elements.

I have grouped the films and TV series in this section according to the different scenarios they envision, which also reflect the contemporary imaginary of catastrophe. The first scenario sees a return to sovereignty and disciplining *dispositifs* governed by the principle of *mors tua, vita mea*, or 'war of all against all'. Iconised in popular culture through alien films such as *War of the Worlds* (1953; 2005), this scenario underpins the ideology of a 'war against Evil' in which the 'saviours of civilisation' join forces against the threat – as described by Simmel (1908) and Elias Canetti (1960) in their seminal works on society, crowds and power – deploying Anthropocenic techno-science to save the world. The second scenario sees the use of artificial selection and genetic manipulation to guarantee the reproduction of a 'superior', 'enhanced' ruling elite. Finally, in the third scenario, biological and political transformation is fostered by trans-cultural/trans-human communication (e.g. *Arrival* and *Annihilation* in Section 2) as well as by lasting cultural and biological contamination.

Chapter 5 discusses the films in the first group – all of them planetary-cataclysm blockbusters such as *Armageddon* (1998), *Deep Impact* (1998), and the more recent *The Wandering Earth* (2019), in which the crisis is over once the threat is apparently defused either by blowing up Earth-bound meteors and comets or by leaving the solar system. The mechanism of temporary solidarity among the ones who possess knowledge and technology that promise a return to the *world as the 'we' knows and wants it* is a staple in all these films, in which First World techno-science, just like Rousseau's 'modern reason' in the case of the Lisbon earthquake, is believed to offer a universally valid solution. In *The Impossible* (Chapter 6) the crisis triggered by the 2004 tsunami in Southeast Asia is overcome through a cell-phone and mutual support among Western tourists. The wish to return to the Anthropocene and its dualism between worthy and unworthy indicates denial of the catastrophe itself, which, similarly to the debate on climate change, is merely regarded as a *crisis* providing a chance for *renewal*, validating the belief that the Anthropocenic is the best possible world. Even reducing carbon emissions and fossil fuel extraction is more about 'hegemonic common sense' (Goodman and Salleh 2013) and readjustments to the Modern Economic Growth narrative (Barca 2011) than dissent. As Barca argues, this 'common sense' or master narrative obliterates 'the social and ecological costs associated with fossil capital, [...] backgrounds the agency of the non-master subjects, and considers their sacrifice as inevitable and necessary to global historical progress' (Barca 2020).

The second scenario, premised on the dualism between saviours and survivors, is best represented by the Brazilian TV series *3%* and the films *Code 46* and *The Impossible*, which I will analyse in greater detail in Chapter 6. In the name of security and prosperity for all, the 99 per cent (Lohan et al. 2011) – the vast majority of humanity – is lumped together into an undifferentiated mass of people whose lives depend on ontologies, logics and decisions imposed upon them by the chosen few, although the reality of their disadvantaged condition is rarely acknowledged. In fact, a progressivist, paternalistic, civilising narrative prevails. All these films (except *Snowpiercer*), paint the masses of the 'drowned' as being on the receiving end of the saviours' benevolence rather than unfair treatment. As such, they are not expected to object to inequality nor are they allowed to jeopardise the superiority of the One World and its rule. Legacies of social Darwinism, eugenics, genocide and biopolitics from colonial rule are reproduced in this scenario and projected onto an apocalyptic event in which human selection and enhancement seem to be the only means to the survival of at least part of *the world as the 'we' knows and wants it*. In the deeply classed, gendered and racialised world to be preserved, only the superior or those who can be further 'enhanced' are deemed worthy of surviving. Social Darwinism is clearly deployed in *3%*, not to mention *The Impossible*, where a white family returns to civilisation leaving behind the destruction and death caused by the 2004 tsunami. *Blindness* offers a similar narrative, in which the survivors are the wealthier, or the ones they befriend, and fittest. *Code 46* is clearly about eugenics, as is *I Am Mother*: Genetic modification and embryo freezing are direct expressions of a divided society in which only genetically fabricated humans can access the safe space inside – where new plantations and colonies for the fittest are created (*I Am Mother*) – while the less 'fit' are either exterminated (*I Am Mother*) or exposed to death by solar radiation (*Code 46*) in the unsafe space outside. In *Code 46* the global divide between inside and outside clearly evokes a postcolonial world where pollution causing the ozone gap affects the masses of wretched of the earth. In *I Am Mother*, the robot's wish to start humanity anew echoes the post-nuclear scenario of McCarthy's *The Road*, in which a father and son struggle to stay human in a world where cannibalism has become a means of survival.

In different ways, both the first and the second scenario refuse to acknowledge that the environmental catastrophe is a *revelation*. Like Walter Benjamin's (1940) angel of history, the environmental catastrophe looks back into the past and gets to question the idea that the Anthropocene may be the only One World possible; then it looks ahead to the future and opens the space for an alternative post-Anthropocenic society. This alternative is provided in the third scenario, which leaves the door open for humans' return to a feral state that I see as symbolising greater proximity to Earth and a critique of Cartesian rationality. In Chapter 7, I will build my analysis of this scenario on the films *Arrival* and *Annihilation* (Section 2), where aliens and humans share either a new language and perception of time, or a common epistemology and refracted DNA. There, I will consider in greater detail the post-humanist narrative of the ultimate catastrophe in *The Girl with All the Gifts* (2016, hereinafter *The Girl*), in which human knowledge is passed down to mutated children who represent the left-outs of a species that has not been able to understand and deal with an environmental apocalypse. In the film, the apocalypse is caused by a fungus that envelops the human brain, transforming humans into rabid cannibals first, and then into humus. Embryos from infected women's wombs survive and grow to become a new species retaining human intellectual competencies within a bodily bio-structure that recalls humanity's animal-like former stages. Human *geography* and *history* are only temporary and are eventually erased by the apocalypse: the post-human masters of the planet, the mutant children, are still partially human (humanity never disappears!), but time and space cease to be linear and fractured by borders and boundaries, while Anthropocenic dualisms, ontologies and logics give way to trans-speciesist *kinships*.

What all three scenarios seem to share, albeit to a different degree, is the refusal to acknowledge the existence of alternative epistemologies and forms of life, that which Arturo Escobar calls the *pluriverse*. All these films, with no exception, disregard the existing assemblages of new environmental, anti-capitalist and anti-patriarchal fights on the one hand, and Indigenous knowledges and struggles on the other, as well as their non-extractivist relationship with Earth and alternative non-capitalist and anti-colonialist forms of production (Wynter 1971; McKittrick 2013; de la Cadena 2015; Dubois and Casimir 2010; Wright 2018; Davis et al. 2019; Milanez and Pinto 2017).

In the next chapter, I will explore how the Anthropocene's *monstrous geontologies* and the dualisms on which they are based silence the plurality of worlds and 'Other' epistemologies, as well as shaping the 'we's' imaginary of the post-apocalyptic future.

3 Enduring dualisms, monstrous *geontologies*

As the anthropologist Arturo Escobar (2016) writes in *Thinking-feeling with the Earth*:

> Although taken as the common sense understanding of 'the way things are', the One World World [OWW] is the result of particular practices and historical choices. A crucial moment in the emergence of such practices was the Conquest of America, which some consider the point of origin of our current modern/colonial world system (e.g., Mignolo 2000). Perhaps the most central feature of the One-World project has been a twofold ontological divide: a particular way of separating humans from nature (the nature/culture divide); and the distinction and boundary policing between those who function within the OWW from those who insist on other ways of worlding (the colonial divide). These (and many other derivative) dualisms underlie an entire structure of institutions and practices through which the OWW is enacted.
>
> (p. 21)

One-worldliness erases the Other worlds: the Moloch makes them 'absent epistemologically' (Sousa Santos 2014), swallows them, and annihilates them through exploitation and extractivism. This has been clearly rendered in James Cameron's *Avatar* (2009), which narrates life on Pandora, a fictional gas giant's habitable moon in the Alpha Centauri star system where the *geontology* of the Anthropocene does not apply. Human attempts to colonise it to extract the mineral unobtanium, a room-temperature superconductor, threaten the lives of Pandora's indigenous people, the Na'vi – 10-feet tall, blue-skinned, sapient humanoids that live in harmony with nature and worship the mother goddess Eywa.

Superior military knowledge, colonisation and extractivism are the weapons of the Anthropocene in search of new colonies in which to expand its system. As a memory of the future, *Avatar* projects onto the future ontologies and logics that have shaped the so-called Plantationocene and 'racial' Capitalocene since the beginning of Modernity (Moore 2015, 2017, 2018; Haraway, 2015; Haraway et al. 2015: Mitman et al. 2015; Vergès 2017; Yusoff 2018). In *Avatar*, the centuries-long 'American holocaust' (Stannard 1992) that began with the Conquest and westward expansion is rendered into a fairy tale. Looking at the catastrophe (that is, colonisation) from the standpoint of a more benevolent coloniser, *Avatar* reveals the situatedness of the idea of apocalypse, while at the same time projecting the end of the world onto a place outside the boundaries of planet Earth or outside what is deemed to be the worthy portion of the world itself, as if the threat of the apocalypse did not touch the 'we'. As Hartman (1997) argues, empathy towards the target of destruction allows shifting the attention from the specific condition of the colonised (the Na'vi) to the coloniser (the humans), thus generating a double discursive effect: 'we' are victims too, and as such 'we' cannot be responsible for the catastrophe. This also conceals the fact that, among the colonisers, the poorer 99 per cent does not have the same access to resources as the wealthier 1 per cent (as Rousseau also argued; Masters and Kelly 1992, p. 110). In the foreseeable future, the 1 per cent (and a few more) will still find a way to look to the disasters in the Out there with indifference and denial and profit from the logics of the Anthropocene without feeling the urge to upset them. The earthquake in Lisbon, one of the largest and most important cities of the time after London, Paris and Naples, killed an estimated 70,000 people, but that is not the reason why it was a shocking event for Europeans:

> While there were other earthquakes which occurred about the same period, notably in Catania, Sicily, and in Port Royal, Jamaica, both in 1693, those events were somewhat irrelevant, from a European viewpoint, having taken place in distant and exotic places. And most of northern Europe was seismically stable.
>
> (Dynes 2000, p. 98)

Dualisms between places worthy of consideration when struck by disaster and those that are not have been maintained in European and colonial Modern history through many different discursive *dispositifs*. These same *dispositifs* define what is considered an apocalypse and what happens instead for the greater good. Patriarchal violence over women's bodies, knowledge and epistemologies (Federici 2004), chattel slavery, pogroms, as well as colonial genocides and environmental devastation in the Americas, Africa and Asia share a common denominator: the catastrophe was imposed on those same individuals and peoples against whom the colonial master defined his identity and boundaries (Merchant 1980, 1989; Stannard 1992; Diamond 1997; Crosby 1986; Byrd 2011; Pulido 2018; Keys 2000; Friedlander 1997; Davis 2000). A catastrophe acts as a semiotic border between the civilised and the uncivilised, identifying who is expendable and who is not and who will pay the highest price for the unwanted consequences of production, progress, social order and political stability (e.g. biocontamination and desertification) on the basis of a colonial and cultural archive of 'figures of race' (Section 2). The semiotic power of the boundary separating the *worthy* from the *expendable* distinguishes between *bad* progress (that which is in the wrong hands and will contribute to the crisis, e.g. prompting large flows of environmental migrants) from *good* progress (led by the rational, profit-driven and wealthy part of humankind) (Bhambra 2007). Accordingly, it determines who can enter the control room, and who is excluded from the space of *politics*. Indigenous knowledge, resistant practices and resilient behaviours of Afro-descendants, peasants, and poor urban communities were and still are validated *politically* only insofar as they can be converted to practices of value extraction. When they cannot, or stand in the way of exploitation and value extraction as in the case of indigenous farming in the Amazon or the counter-plantation system developed by fugitive slaves, they become targets of epistemic and physical genocide (Swyngedouw and Ernstson 2018; Yusoff 2018, pp. 206–207). In an interview with Arturo Escobar, an Afro-descendant activist from La Toma, in Southwest Colombia, describes the community's long struggle against gold mining: 'it is patently clear to us that we are confronting monsters such as transnational corporations and the State. Yet nobody is willing to leave her/his territory; I might get killed here but I am not leaving' (Escobar 2016, p. 19). This kind of activism is an expression of what Povinelli calls the figure of the 'Animist'. Animists believe that specific features of the landscape such as sacred rocks are sentient and do not accept the *geontological* assumption that Nonlife is not *political* and hence not worthy of respect. The apocalypse is nothing new for people

> who for four centuries had been subjected to the impact of the 'great discoveries', of empires, modernization, development, and finally globalization [...] [as] they knew perfectly well what it meant to find oneself deprived of land. And they even knew quite well what it meant to be chased out of one's land. They had no choice but to become experts on the question of how to survive conquest, extermination, land grabs.
>
> (Latour 2018, p. 21)

Today, disaster management and reconstruction programmes are the primary site of production of a new Western and neocolonial discourse of *worthiness* or *deservedness* (see Butler, 2009, in the frame of the war on terror) – that is, the act of selecting who is 'deserving' or 'undeserving' of aid through the concept of humanitarian precedence. Think of the many disasters that have happened in the last ten years, from the 2010 earthquake in Haiti that affected 3 million people and killed 230,000, to hurricane Idai, which hit Mozambique, Madagascar, Zimbabwe, Malawi and South

Africa in March 2019, causing 800 victims and the destruction of several villages.[9] They have quickly faded from the news and from Western collective memory, as if the victims were less worthy of grieving and their suffering was to be blamed on a specific human ontology that – unable to deal with disaster due to insufficient technology and knowledge, and unequal valuation of human lives – is expected to somehow coexist with disaster.

Besides, if we compare media coverage, public debate and financial aid in the Mozambique flooding, the fire at Notre Dame cathedral in Paris in which no casualties occurred (see also Armiero and Turham 2019, pp. 37–38) and the disasters at sea caused by the border regime in the Mediterranean (with a death toll exceeding 20,000 in the years 2014–2018) (Casid 2018), we can see that the meaning of disaster is subjective. Institutions, the media and public opinion have a different understanding of its size, relevance and consequences that is independent of the underlying data.

A binary rhetoric locating forgettable 'disasters' beyond the borders of the worthy 'imagined community', on the other side of the gate separating visibility from invisibility, differentiates between places *of* disaster (where catastrophe can be controlled and therefore overcome) and places *for* disaster – where disaster is part of people's daily lives. In colonial narratives, the discovered lands were a place *for* disaster in which indigenous people were overwhelmed by natural and human-generated calamities that they were unable to master as they did not have the necessary culture (Giuliani 2017a, p. 76). In the colonial metropole, or in the postcolonial world divided by the iron curtain, disasters occur where technology is lacking or improperly used. As Dynes argues in his analysis of the debate between Voltaire and Rousseau on the Lisbon earthquake, today:

> many of the themes of the Enlightenment persist. As a part of the premium placed on reason, science and technology as the prime 'solutions' to disaster continue to be recommended. Whether it is better construction materials, the building of levees and dams, better detection methods, enhanced communication technology, or a variety of other technological fixes, modern disasters are usually considered primarily technological failures. It is also difficult to compare the level of intensity of the arguments evoked by Lisbon with the level of intensity generated today by Bhopal or Chernobyl. But Lisbon, Bhopal, and Chernobyl all illustrate the contention that the meaning of a disaster is always seen in the cultural context of the time.
>
> (Dynes 2000, pp. 110–111)

Nowadays, as in the past, disaster coincides with a number of phenomena, from wars to tsunamis, from epidemics to inhuman labour exploitation: in brief, global inequality in its most extreme form. As Voltaire, Rousseau and more recently Mike Davis (1990; 1998; 2002; 2005) remind us, the binary opposition between worthiness and deservedness – and the distinction it operates – does not only apply to the Out there, the postcolonial elsewhere. Places *for* disaster can also be found (or built) within the Global North, as Davis has shown in his research on the effects of nuclear and chemical testing, urbanisation, and desertification in the United States since the aftermath of the Second World War. More recent examples quoted by Dynes include the Bhopal (1984)[10] and Chernobyl (1986) disasters. In Bhopal, the Indian government allowed Union Carbide, a U.S. pesticide manufacturer, to prioritise profit over safety, resulting in a gas leak that killed at least 8,000 (Lapierre 2002). In the case of Chernobyl, the cost of poor management by the U.S.S.R. central government will be paid by generations of Ukrainians, as radioactive contamination is expected to last for 300 years (Abbott 2006).[11] Similar incidents in Italy involved collusion between state institutions, polluting factories and local governments in the industrial areas of Naples, Seveso, Taranto, Marghera and Manfredonia, and their careless management of environmental and health disasters caused by chemical accidents and toxic waste (Armiero and de Rosa 2016; Armiero and Fava 2016; Centemeri 2006; Barca and Leonardi 2016; Zazzara 2009; Malavasi 2018). Lack of consideration of environmental factors caused the 1963 landslide of the Vajont Dam that killed nearly 2,000 people (Armiero 2011).

The definition of disaster depends then on many local and global factors and layers of power relations. Generally, it is not provided by the disaster-struck community, which does not have a say in short- or long-term disaster management either. (In most cases, disaster is defined by actors who are in the driver's seat but are not directly involved.) Disaster is seemingly not seen in its complexity as the result of the interaction between human, non-human, organic and inorganic agents. In this thoroughly other-directed meaning of disaster, what is lost is what 'natureculture' feminism calls multifactoriality, *trans-corporeality* and *intra-activity* of natural (physical and biological) phenomena (Tuana 2008; Alaimo 2008; Barad 2007, 2008), that is to say, the internal interactions of multiple factors that contribute to generating a given phenomenon. In agentic materiality, a material reality in which all the elements and factors involved in creating phenomena are viewed as 'agents', human beings are neither the sole or main agentive element nor the blameless victims. Not even Hurricane Katrina can be termed a natural phenomenon independent of human intervention, since humans are an integral part of the natural environment (Tuana 2008; Clark 2011, pp. 140–143). They are 'acted-on agents' who can influence and change the physical contexts in which they live while, at the same time, being influenced and changed by them. When viewed from this perspective, the narrative that explains the December 2004 tsunami in Southeast Asia, the tidal wave that broke over Fukushima in March 2011, the wildfires that devastated Australia in late 2019, or the 2020 coronavirus pandemic as being caused exclusively by 'natural factors' (neglecting to acknowledge the role played by global warming and the effects of human behaviour on the environment) is not particularly plausible (Clark 2011, p. 63). The dominant rhetoric that casts humans as the only actors capable of solving things is unacceptable. Neither can we assume that the subjects affected by a disaster would be unable to survive in harmony with (some of the) climate changes and develop new forms of life that diverge from the 'logics' of disaster management and humanitarian intervention.

Specifically, negating the subjectivity of the people affected by disaster casts the non-West as a homogenous space that is 'naturally defined' as the place where disasters occur, in line with the definition of places *of* disaster and places *for* disaster that I propose here. In this way, it could be argued that the Out there is once again positioned as a 'colonial object', in which disaster is a frame rather than an event (Giuliani 2016a; 2017b, p. 229). As such, it must be subjected to strict military, political and governmental control through humanitarian intervention or, when it poses a security concern, through the 'small wars' described by Asad (2007).

When this spatially specific and reassuring dichotomisation between places *of* disaster and places *for* disaster is overturned, disaster will tip over the borders into the so-called Global North, bringing with it inequality and 'chaos'. It will not be a regenerating catastrophe but an absolute disaster, the end of the *world as the 'we' knows it*. The 'we's' most pressing fear is that the actual future will not look like the imagined future. *History* and *geography as the we knew them* are vanishing, leaving behind the monstrous threat of disaster: epidemics, destruction, human genetic modification and wars without borders, potentially limitless in time and scope, are a common theme in many post-apocalyptic films (*Blindness, I Am Mother, The Girl*), some of which envision a safe space where humankind will have the opportunity to start anew (*The Impossible, Elysium, 3%, Blindness*). Despite the threat not being over and safety being precarious, no alternative way of dealing with the crisis is envisioned for the time being. As we will see in Chapters 5, 6 and 7, the crisis is believed to be averted in all the films and TV series I include in my analysis except *The Girl*, where a lasting mutation is felt as ineluctable. Next, I will briefly outline the controversial debate surrounding the Anthropocene and the alternative modes of thinking it has generated.

4 Fighting within and against the monstrous Anthropocene

In *Geontologies. A Requiem*, Elisabeth Povinelli argues that in Aboriginal cosmology:

> we could distinguish between [...] two forms of material graphing as *in sutu* rather than in situ, respectively, a perspective that emphasizes a given or changing suturing that creates various modes of existence and a perspective that emphasizes the various modes of existence in the

> situation. How should one conceive manifestations that alter one's understandings of the *in sutu* rather than merely perceive the elements within a given assemblage (the verb *stem gumen*, to manifest, versus *gaden*, to see)?
>
> (2016, p. 90)

In Sections 1 and 2 I have explored the *suturing* of Western *history* and *geography* that accompanies human mobility from the Global South to Europe and the West. I have shown how this process erases the epistemological separation between a forgotten history of horrors and a supposed violence-free present, and that between distanced places ridden with violence (the Out there) and the civilised place of the 'we' (the Here), threatening the 'we' that through *geography* and *history* has built its own superiority and safety. The operation of *suturing* concerns a situated perspective that assembles elements of *bios*, *zoe*, and *geos* in a temporality that connects past, present and future. As in the case of the suturing processes that I have described in Sections 1 and 2, Aboriginal situatedness 'presupposed that they lived in a world with multiple involvements and co-constitutions, all of which could be rearranged to the benefit or detriment of each part' (Povinelli, p. 92). This implies a vision of Earth in which the *partage* described by Iveković (2019) supersedes Anthropocenic *geontologies* to include all organic and inorganic organisms, which explains the Aboriginal claim that the Two Women Sitting Down rock formation is sentient. It also explains why Aboriginal resistance to drilling and extractivism is more about non-anthropocentric living experiences and epistemologies based on the idea that human life is enmeshed into a 'web of life' than traditional land protection (Danowski and Viveiros de Castro 2017). Their claim is echoed by many similar struggles across the global North–South divide and in spite of the multiplication of borders for labour exploitation and value extraction – e.g. protests against a high-speed rail link in Northern Italy or the Dakota Access Pipeline on Meskwaki and Sioux lands.

These are but a few examples of *suturing* in the panoply of acts of resilience and resistance to the destructive effects of the Anthropocene claiming for *pluriverse* to be acknowledged as viable *political* alternatives. Such acts of resilience are reshaping the spaces of *the political* and express new modes of dealing with the climate crisis, reflecting a *political* discourse on the environmental catastrophe that incorporates five centuries of indigenous resistance against land grabbing and fossil extractivism as well as grassroots action against climate change. They also coincide with a broader radical change in social and agroecosystems – no longer based on fossil extraction, overproduction, animal and labour exploitation (Córdoba Vargas et al. 2019). Indeed, the policies designed by the 'we' are deemed to be nothing but palliatives and the Global North CO_2 reduction agreements, popularised under the name of *green economy*, are revealed as nothing more than a 'capitalist attempt to incorporate the environmental limit as a new terrain for accumulation and valorization' (Leonardi 2017, p. 171). *Biomimicry* (Benyus 1997; Dicks 2016), according to which productive systems need to be made into an 'imitation of natural models, systems and elements to reduce negative environmental impacts of productive activities without sacrificing economic growth' (Leonardi 2017, p. 171), is also regarded as merely a neoliberal reconfiguration of governmental *dispositifs* and biopolitics.

These forms of struggle in dialogue with degrowth movements, post-development theories and material feminisms are calling for deceleration and alternative systems of social production (Kothari et al. 2019). Degrowth, the feasible utopian plan theorised by André Gorz (2010), aims at rebalancing the relationship between humans, non-human life and non-life by slowing down the pace of capitalism and reducing alienation (Barca 2017). What degrowth means, then, is not only the decreased growth of existing production systems but more importantly the creation of new ones – 'our emphasis here is on *different*, not only *less*' (Kallis et al. 2015, p. 4) As Emanuele Leonardi explains, Gorz's *political ecology*:

> *is a form of anti-capitalism*, an activist-framed analysis of potential cracks within the logic of capital, a utopian effort to actualize—here and now—practices that unhinge the system of techno-economic compatibilities. [...] To envisage a different economy, different social relations, different modes and means of production, and different ways of life is regarded as 'unrealistic', as though the society based on commodities, wages, and money could not be surpassed. In reality, a whole host of convergent indices suggest that *the surpassing of that society is already under way*, and that the chances of a civilized exit from capitalism depend primarily on our capacity to discern the trends and practices that herald its possibility.
>
> (Leonardi 2019, p. 60)

Degrowth and post-development theories need to be accompanied by a critique of the integrated processes of production of goods and services and reproduction of labour power in the Anthropocene, both within and outside capitalism. Some eco-feminist reflections provide critical ontologies of capitalist production and forms of alienation based on the valorisation of what Stefania Barca has called 'forces of reproduction':

> This concept is an attempt at producing a new, powerful conceptual tool that may make visible and support the struggles that take place around the work of earth-care. In a rather a-systematic way, the concept reflects the influence of two distinct theoretical traditions: ecofeminist thought and historical materialism. A point of departure here is Social Reproduction Theory (SRT), and particularly Nancy Fraser's characterization of capitalism as an expanded system, not

> simply economic but social, political and ecological, with an 'inherent but blind directionality', a 'self-expansionary process through which it constitutes itself as the subject of history, displacing the human beings who have made it and turning them into its servants'.
>
> (2020 forthcoming)

Social reproduction, grounded in practices of care and earth-care, is seen here as the alternative *modus cogitandi, vivendi* and *operandi* that promises to suture global social divides and reverse Anthropocenic *geontologies*. Social reproduction, which is at the core of Maria Mies' (1986), Silvia Federici's (2004) and Nancy Fraser's (2014; see also Arruzza et al. 2019) feminist political stance, presupposes degrowth as the solution to exploitation and extractivism, as opposed to neoliberal imperatives of production and capital's 'inherent but blind directionality' (Fraser 2014). It proposes a living epistemology in dialogue with native and alternative knowledges and practices on the basis of a feminist understanding of 'care'. Yet, as the feminist critique of reproduction has shown in the last four decades, care is neither socially neutral nor essentially a positive concept. Reproduction is at the root of a system of exploitation that since capitalist and colonial Modernity has been violently imposed on gendered individuals – in line with the sexual division of labour, essentialised gender roles, and normative ideas of femininity and masculinity (Lugones 2011; Walsh 2015). The reproduction of labour, and enslaved labour, was imposed on women and other gendered subjects along and across the colour line, forcing women into that 'traditional compulsive altruism', as Patricia McFadden (2020 unpublished) calls it, which reduced their and other gendered subjects' autonomy, as well as jeopardising their physical, psychological and intellectual integrity. (Forced) care was not only a system of oppression, but also a necessary *dispositif* to reproduce labour exploitation, racialised objectification, the genocide of indigenous peoples and devastation of the planet. Reproduction and care are thus not signs but signifiers, which feminist critique of anthropocentrism ought to resignify in accordance with critical ideas of bodily and moral integrity, sustainability, and accountability of knowledge production and social action (Tola 2018).

In fact, even a robot may be tasked with social reproduction and be capable of raising a generation of enhanced humans after exterminating those who did not meet the highest standards of responsibility and solidarity (*I Am Mother*). The robot has a higher conception of humanness, one that is, nevertheless, is still grounded in the imperatives of anthropocentric rule over nature.

A different idea of care – a frightening one because it subverts gender roles, class hierarchies and colour lines – is based on a postcolonial epistemology of presence (Sousa Santos 2014) and acknowledges the political importance of accountable affects and proximity. Sara Ahmed's (2000) *skin-to-skin* proximity, Iveković's (2019) *partage* and Escobar's (2016) pluriverse sustain and further articulate a critique of the Anthropocene and its *geontologies*. Films such as *Annihilation* (Section 2), *Arrival* and *In the Flesh* (Section 1) add another element to the picture: political acknowledgement of the interdependency of Life (social and organic) and Nonlife. Acknowledging that humans do not merely 'forge connections' with the Earth and nature but are 'utterly dependent' (Clark 2011, p. 30) on them is a political statement that underscores reciprocity, or *trans-corporeality*. Its consequence is the repoliticisation of the *strange encounter* (Ahmed 2000) as a harbinger of new life and hope. And, as Borghi (2014) reminds us, the political value of accountability within a system 'with multiple involvements and co-constitutions, all of which could be rearranged to the benefit or detriment of each part' (Povinelli 2016, p. 92) comes from acknowledging the political value of *trans-corporeality* and *intra-activity*. Then again, attaching political value to *trans-corporeality* and *intra-activity* means acknowledging the disparity in power between different agents in an integrated system where differential (geo)power relations are recognised but cannot be erased by a politics of location. Humans, then, will have different transformative power over the environment than other actors, such as water or viruses. In any case, my argument stays in line with Elizabeth Grosz (2011, p. 71):

> Unlike Foucault's freedom, where self-care (of which the most salient element is *parrhesia*) becomes the foundation for freedom (Fornet-Bettancourt et al. 1987; Foucault 2008 [1984]), Grosz argues that it is not subjects who are or are not free, but actions undertaken by living entities as they relate to other living entities and materiality that foster the 'ability to make (or refuse to make) activity (including language, that is, systems of representation and value)'.
>
> (Lloro-Bidart and Semenko 2017, p. 21)

This system is based on what radical feminism, and more specifically African and African American radical and lesbian feminisms, identify as the necessary tension between care, earth-care and self-care: as subjects of responsible action immersed in a system of *trans-corporeal intra-actions*, humans should not only take care of Life and Nonlife but also preserve their autonomy, integrity and pleasure. In fact, self-care – meaning personal autonomy and independence, integrity and sustainability – is an essential prerequisite to stripping care of its patriarchal significations. For 'care' to be turned into a political practice, it needs to be resignified away from its traditional association with essentialised (sexualised, gendered, classed and racialised) conceptions of reproduction – as care cannot be separated from self-care and self-love.

In her epilogue to the 1988 collection of essays titled *A Burst of Light*, Audre Lorde wrote that 'caring for myself is not self-indulgence, it is self-preservation, and that is an act of political warfare' (1988, p. 227). As Sara Ahmed explains, Lorde is making a crucial point: for those who are driven to compulsive care-giving, self-care is an act of political warfare rather than a technology of power (Foucault 1976) because 'in directing our care towards ourselves we are redirecting care away from its proper objects' (Ahmed 2014).

> And that is why in queer, feminist and anti-racist work self-care is about the creation of community, fragile communities, assembled out of the experiences of being shattered. We reassemble ourselves through the ordinary, everyday and often painstaking work of looking after our-selves; looking after each other. This is why when we have to insist, I matter, we matter, we are transforming what matters.
>
> (Ahmed, 25 August 2014)

Fragile communities are those that include vulnerable subjects. In Inna Michaeli's words, 'we are not talking about a generic "self" here, an abstract individual, but about a self which is grounded in particular political histories and present situations of violence and vulnerability' (2017, p. 53). These are communities of racialised, cisgender and transgender women and queer people who struggle to ground their lives in self-care, self-love and pleasure, as well as in non-normative and non-reproductive sexuality.[12] 'Self-care' – argues Ahmed – is 'not about self-indulgence, but self-preservation [...]. This kind of self-care is not about one's own happiness. It is about finding ways to exist in a world that is diminishing' (Ahmed 2014). Self-care practices are at the core of earth-care. In fact, earth-care, pleasure, eroticism and non-reproductive sexuality are essential features of the Chthulucene (Haraway 2016, p. 2). Conceived within a *trans-corporeal* system of accountability and mutual implications (Alaimo 2008, 2010, 2016, 2018) that precede *intra-actively* the formation of subjects and objects (Barad 2008), self-love and care as well as growing one's own food become politically meaningful and subversive acts within a process of conscious subjectivation that seeks to avoid egoism, exploitation and irresponsibility. And to avoid the 'neoliberal trap' (Michaeli 2017) of a privatised, depoliticised, consumerist conception of 'self-care', the latter cannot be separated from care.

Both my *political critique* of the Anthropocene and my *alternative* political project are grounded in the indissoluble interdependency of care, self-care and earth-care. As such, both aim at the disintegration of Anthropocenic patriarchy – a necessary prerequisite to the end of capitalist systems of endless value extraction.

Unlike the fantasy of Anthropocenic *renewal* described in Chapter 2, *trans-corporeal* and *intra-active* epistemologies suggest alternative modes of individual and social life based on a break with privilege-producing *dispositifs*. At the same time, they *reveal* the *monstrous* and evil nature of racialised, gendered and classed *dispositifs* of objectivation at the core of humankind's unlimited capacity to pollute and destroy while blaming monsters of its own making.

They also reveal the partiality and situatedness of past and present processes of monstrification (Foucault 1961) against forms of life that Western societies have threatened: the *monsters* of Western Modernity were nothing but the result of multiple caesurae between past, present and future and between the Here and the Out there.

In the next chapters, I will show how films and TV series are symptomatic of the unwillingness in mainstream and popular culture to envisage a radical upsetting of the Anthropocene, which generally results in no more than mild criticism of the 'side effects' of capitalism. *The Girl with All the Gifts* (2016), which I will discuss in Chapter 7, is the only exception.

5 The (white) saviour of civilisation

In Jon Amiel's *The Core* (2003), Dr Josh Keyes (Aaron Eckhart), a geophysicist, discovers that the Earth's magnetic field is getting increasingly unstable when strange events are reported around the world. Together with two scientists and astronauts from the Space Shuttle *Endeavour*, he will reach the Earth's core to set off nuclear bombs and restart its rotation. Only two crew members, the astronaut Rebecca Childs (Hilary Swank) and Josh, will survive and return home safely. A sort of apocalyptic version of Jules Verne's *Journey to the Centre of the Earth* (1864), the unlikely plot centres around a mission to prevent Earth from turning into a giant ball of fire with the aid, somewhat paradoxically, of the deadliest weapons ever created. The planet's magnetic field has started deteriorating following a failed top-secret project called 'DESTINI' (Deep Earth Seismic Trigger INItiative) that was expected to generate earthquakes to maintain the core's rotation and instead caused its slowdown. As in *World War Z* (2013) and Francis Lawrence's *I am Legend* (2007, see Section 1), humankind is faced with the consequences of mistakes made in the name of civilisation, but its faith in techno-science never waivers. On the contrary, once again its survival is made to depend on destruction, as the same techno-scientific progress that led to a catastrophe seems to be the only way out of it. The use of nuclear bombs to prevent Earth's destruction was first introduced in two classic apocalyptic films, *Armageddon* (1998) and *Deep Impact* (1998), where the (white) saviour of civilisation is hailed as a messiah willing to sacrifice himself for the good of the world.

In Frant Gwo's *The Wandering Earth* (2019), a planetary collision is threatening the Earth. In the year 2061, the aging sun is about to turn into a red giant that threatens to engulf the Earth's orbit within 300 years, forcing the nations of the world to initiate a project to migrate the Earth out of the solar system to the Alpha Centauri system. While most of the Earth's surface is already frozen, forcing the severely reduced human population to live underground, the military and a team of astronauts try to avert the catastrophe with the help of huge planetary thrusters. After a few failed attempts, the Chinese astronaut Liu Peiqiang (Wu Jing) finally succeeds in igniting Jupiter's atmosphere and saves Earth from destruction. In the last scene, his son Liu Qi (Qu Chuxiao) describes the stages of the Earth's migration, a process that will take 2,500 years and 100 generations to complete.

In both *The Core* and *The Wandering Earth*, the apocalypse propels fantasies of a renewed humanitarianism aimed at saving as many lives as possible regardless of hierarchies of class, gender, race, nationality and sexuality, and regardless of the memory of the horrors of colonisation. These hierarchies are concealed in the film, signalling the uneasiness of mainstream cinema about the impact of power and capitalism on the vulnerable. As Saidyia Hartman (1997) has argued, empathy towards victims and survivors of a catastrophe shifts the attention from the vulnerable to the hegemonic subjects, who are therefore constructed as representative of all humanity. The neutralisation of ethical and political issues related to selection, life extermination and colonisation turns *The Core* and *The Wandering Earth*, among others, into narratives celebrating the U.S. as the saviour of the world thanks to its techno-science, and China as able to match its technological capacity. The whitening of the saviour is the corollary of a universalistic outlook that sees this approach and methods as the only possible and effective solution. Ideological unity among humans is fostered by fantasies of a brand new start and nurtures the imagination of a father who will do anything to protect his children and the whole humankind. From his viewpoint, the catastrophe has turned the Earth's population into a needy child waiting for a saviour to come to the rescue, a benevolent and loving father/civilisation/nation that takes care of everything and everybody, when instead it is a very limited portion of humanity that he is seen as taking care of.

In *The Day After Tomorrow* (2004), Jack Hall (Dennis Quaid), a paleoclimatologist, tries to warn the world of an impending climate shift. When he presents his findings at an international conference in India, however, he encounters fierce opposition from the U.S. state secretary, who is more interested in the global energy crisis than in the effects of global warming. Jack explains that the rapid melting of the glaciers will change the equilibrium between salted and sweet waters, which in turn will affect the Gulf Stream's capacity to maintain warm temperatures in the northern hemisphere, leading to a new glaciation within a few centuries. Contrary to his expectations, the effects start appearing in less than 48 hours, and a climatic apocalypse transforms *the world as the 'we' knew it*. Jack's teenage son, Sam (Jake Gyllenhaal), is in New York, where he finds shelter at the National Library. Meanwhile, Jack manages to persuade the president of the United States to at least order the evacuation of the southern states, as the northern ones are already impossible to reach. He is finally able to join Sam and other survivors after the storm has reached New York and glaciation has started. In the last scenes, the military rescue all the survivors who had found shelter inside Manhattan's buildings. Climate refugees from the United States try to trespass the Mexican border, but they will be let in only after the newly appointed U.S. president cancels Latin American debt. Despite this debt being the legacy of colonial and neocolonial regimes and ruthless forms of exploitation and extractivism (Lewis and Maslin 2015; Davis and Todd 2017), the *monstrous* racialised South is seemingly offered a chance at 'redeeming' itself by rescuing members of a superior civilisation. Unlike in *Ice 2020* (2011), border violence – that same violence that the United States and Europe are inflicting on migrants and climate-change refugees – is not visible in *The Day After Tomorrow*, as if climate-change induced migration from the North to the South were a legitimate cause for human mobility and current migration to the North were not. Both are caused by the same inequality and economic processes that are at the root of climate change, but a distinction is made between the two, based on the tacit assumption that the Global South is dependent on the Global North. Mass mobility from the South is regarded as an ordinary event. Mass mobility from the rich North towards the South is instead deemed to be *extraordinary*. At the same time, this discourse naturalises South–North mobility as *exceptional* in contrast to a view of the North as 'settled, sedentary and at some degree removed from the transnational flows of labour' (Baldwin and Bettini 2017, p. 3). As Baldwin and Bettini argue, the relation between climate change and migration is, in fact, fundamentally *political*:

> We argue that the *relation between climate change and human migration* must be understood foremost as a relation of power rather than as a hard fact awaiting to be discovered, or as an empirically observable phenomenon. [...] In its more hegemonic expressions, it orients publics to climate change in a way that reinforces the exceptional status of the migrants or refugee. It demands that 'we' view the migrant and the refugee as the 'other', the constitutive outside or excess of what is otherwise imagined as the pairing of 'normal' even if fraught geopolitical and climatic conditions.
>
> (2017, pp. 2–3)

Furthermore, in *The Day After Tomorrow* as in *The Core* and *The Wandering Earth*, there is no mention of how racism, sexism and classism affect people's ability to survive the catastrophe: here, like in the other films, the (white) saviour of civilisation will rescue his children and that will be sufficient, as if their survival were guaranteeing the preservation of the highest human values.

Referring to *The Day After Tomorrow* and *The Road* (by John Hillcoat, UK 2009, based on McCarthy's novel), Andrew Baldwin has argued that both films are 'suggestive of a cultural politics of "climate change and migration" in which the heteronormative family form becomes the privileged economy of desire for surviving climate change' (2017, p. 211). I agree with Baldwin, although I believe that the figure of the patriarch, in particular, and his relationship with his children are pivotal here (the mother, as Baldwin notices, is almost always absent). All these films, except *The Wandering Earth*, are deeply entrenched in white American popular culture and *topoi*, one of which is the foundational relationship between the heroic father and his son. Similarly to other U.S. post-apocalyptic films (e.g. Darabond's *The Walking Dead* and Forster's *World War Z* in Section 1 and Spielberg's *War of the Worlds* in Section 2), the white male hero, who is always a scientist or former military, is the quintessential embodiment of patriarchal love, responsibility and abnegation. As a leader, he commands respect despite his flaws and vulnerabilities, which also fulfils the script of the genre.

In general, all these films and TV series share a view of catastrophe as a means to *renewal*, a rebirth of anthropocentrism grounded in patriarchal and heteronormative love relationships and military order that guarantee social and political stability. They also share a view of the 'we's' techno-scientific knowledge as superior – and the only means for facing up to the post-human epidemic/environmental apocalypse. Finally, they all envision a post-apocalyptic society built on the same foundations that supported the plantation and the walls of settler colonialism – and this is especially true of *The Wandering Earth*, *The Day After Tomorrow*, *The Walking Dead* and *World War Z*.

South Korean director Bong Joon-ho's dystopian film *Snowpiercer* (2013), based on Jacques Lob, Benjamin Legrand and Jean-Marc Rochette's graphic novel *Le Transperceneige*, challenges these fantasies. In 2031, a high-speed train doomed to circumnavigate the Earth is carrying the only survivors of a new ice age that began 17 years earlier. The train carriages, divided along class lines, reproduce the structural violence of neoliberalism: in the slum-like tail section, people survive in miserable conditions, eating protein bars made of insects, with very little water and no medical care. Their children are kidnapped by armed guards to work in the engine room. Any resistance is violently quashed, and torture is a daily occurrence. Nonetheless, the passengers in the front cars look up to the train's owner, Wilford (Ed Harris), the 'benevolent' saviour of the last humans on Earth whose actions cannot be questioned.

The central plot element is a revolt in which passengers from the tail section try to reach the head of the train, kill Wilford, and impose a just order. The rebels reach the train prison and free a security engineer, Namgoong (Song Kang-ho), and his clairvoyant daughter Yona (Go Ah-sung), in hopes of persuading him to open the carriage doors he designed. Namgoong was detained for being a drug addict, and he plans to use 'kronole', a drug that doubles as a potent explosive, to blow the door to the outside, believing they can all survive the explosion. After losing most of his comrades to the armed guards, Curtis (Chris Evans), the revolt's leader, reaches Wilford's suite. He refuses to light Namgoong's explosive, because he still thinks that the train is the only place on Earth where humans can survive. He then goes on to explain that when people first boarded the train, they were left for months with no food or water and the tail section turned to cannibalism. Before killing Wilford, he wants to know the reason for such violence and deprivation. Wilford tries to convince him that he has done the only possible thing, because military order and class divisions are the only *dispositifs* that may prevent a state of *homo homini lupus*. He also tries to persuade Curtis that all revolts on the train have always been planned by him and are nothing more than a Malthusian device to control demography and make life sustainable. Curtis is about to lose his determination and accept Wilford's offer to become the new conductor when he discovers that a friend's son, five-year-old Timmy, has been kidnapped to keep the engine on. After rescuing the child, he stops the engine and gives Yona a match to ignite the kronole. The noise from the explosion causes an avalanche that derails the train. Yona and Timmy flee the wreckage and see a polar bear in the distance, confirming Namgoong's intuition that life exists outside the train. The *master*, as Barca (2020) would call Wilford building on feminist ecologist Val Plumwood's work (1993; 2002), is revealed in all his evilness: his false benevolence was a cover-up for the eternal yoke imposed on the poorest and most vulnerable. Violence, torture, totalitarianism, class hatred, and the celebration of techno-scientific supremacy have an ideological basis and produce inequality. *Renewal* is epitomised in the last scene by the two 'special' children surviving the crash – a liberated slave and a clairvoyant.

6 Surviving the catastrophe: human selection, *renewal* and genetic modification

Fantasies of improving human societies by reducing or discarding those elements that represent the mass of the wretched of the Earth in the white noble and bourgeois mind are as ingrained in the nature of the Anthropocene as is the bodily and intellectual enhancement of the fittest. The idea of *renewal* through violence has a long history, since wars and genocides have always been thought of as means of renovating warriors' and soldiers' virility through murder. In Western pre-modern and modern history, epidemics, too, were thought of as means of reducing social pressure. With the Enlightenment came the idea that, in order to control the risk of individual or societal degenerative metamorphosis, human ranks should be separated and 'improved' through social organisation and medical intervention. Nineteenth-century Malthusianism – thus called after the English economist Thomas R. Malthus (1766–1834) – rationalised the human urge to 'select', calling for the state to reduce the rate of growth of the poorer class. Yet Malthusianism was not 'a sign of the times' but rather an expression of class hatred combined with the modern obsession for state control over the population (Foucault 1976). As such it has resurfaced in the last two centuries within a conception of society that in the West has been justified both in terms of sustainability and in terms of a needed *renewal* of national and human society (see also Harvey 1974; Peet and Watts 2004; Peet et al. 2010). According to Malthus, sustainability and the improvement of national society were possible only provided that the great mass of the working class and the lumpenproletariat were contained. Emigration and unwilling confinement to institutions were also employed in the past as measures to rid of 'inferior' subjects by sending them 'elsewhere'. Social Darwinism (whose main proponent was the English philosopher Herbert Spencer, 1820–1903) was imbued with the same principles: sustainability (for the rich), and *renewal* and improvement (for the élite) through selection (of the poor). There is still widespread appreciation of those principles today, as evident from Angela Merkel's selection of the better educated Syrian refugees (2015–2016), Australia's immigration policies that for years have barred entry to unskilled and HIV-positive immigrants, and the U.S. government's decision in *Deep Impact* to shelter only a privileged few in subterranean galleries (the Arc). Active and violent state intervention in times of peace has become, since Spencer's and Francis Galton's (the inventor of eugenics, 1822–1911) times, a way of achieving artificial selection through health and immigration policies, spatial segregation, and differential access to education, childcare, and other resources. All those measures were somehow connected to Galton's principles of *struggle for life and death* and survival of the fittest. More radical measures aimed at the 'improvement' of society, such as selective sterilisation and the gassing of the disabled, were implemented in the first half of the twentieth century respectively in Sweden, Great Britain, Australia and parts of the United States, and in Germany. They amounted to what Roberto Esposito (2004, p. 11, p. 110ff.) has described as a 'surgical' approach to the 'disposal' of inferior subjects,[13] based on the idea that there is a pathogenic socio-racial element with the power to corrupt the allegedly racially homogenous social body, and that only by 'surgically' eliminating it can the continued purity/racial supremacy of the nation be ensured. If we project this approach onto a global stage and consider it more broadly as the motivation behind moves to select among the socio-racial components of the so-called Global North, we can find its traces even in today's humanitarian aid schemes for disaster victims. The films I analyse in this chapter provide clear examples of this. Given the impossibility of openly calling for a return to 'negative eugenics' of the past, a dystopian version of neo-Malthusianism (Hardin 1968; Boulding 1966) sees natural disaster as a device offering alternative means of selecting among 'inferior' subjects so as to draw more rigid 'naturalised' boundaries between class, gender and race and 'preserve' the sites of power. Such was the case when Hurricane Katrina hit with devastating effects the coasts of the Gulf of Mexico, Mississippi, Alabama and Louisiana in 2005, as well as destroying the historical district of New Orleans.[14] The U.S. government's slow response and insufficient preventative measures, a consequence of the country's long history of discriminating between lives worthy/unworthy of saving along class and colour lines, became a source of national shame.[15] In the context of the threats posed by climate change, mass migration and terrorism, fantasies of renovation, sustainability by selection, and human enhancement often combine to make the post-apocalyptic world more suitable to the fittest and racially superior few (Baldwin and Bettini 2017, pp. 7–9).

The belief that humans evolve and select 'naturally', and that technology merely accelerates or increases human capacity for enhancement, is the result of both capitalist maximisation of value-extraction and authoritarian fantasies of genetic superiority over Others. While 'human controlled selection' and 'bodily and intellectual enhancement' are regarded as positive, unruly selection and mutation in the master's body and society – which entail a significant loss of control for the master – have always been regarded as dreadful. In the 'we's' eyes, Malthusianism, human selection and unruly metamorphoses affecting 'the few' are the worst possible nightmare. The films and TV series that I discuss in this chapter, namely *The Impossible*, *Blindness*, *Code 46*, *I Am Mother* and the Brazilian web television series *3%*, are the expression of different fantasies and nightmares about the climate, environmental, and technological apocalypse.

I have discussed Bayona's *The Impossible* in some of my previous works (2016a, 2020) focusing on the (post)colonial divide between places *of*

Figure 3.2 New Orleans underwater in the aftermath of Hurricane Katrina, 11 September 2005. https://pixabay.com/photos/people-outdoors-industry-house-3294653/.

disaster and places *for* disaster. The film narrates the true story of a Spanish family that lived through the 2004 tsunami while on holiday at a Thai resort (although their nationality has been changed from Spanish to British). The title conveys exactly what I seek to address here: white, Western, middle class, 'respectable' individuals are not supposed to fall victim to disaster. Both in the film and in real life, that fate is reserved for the Thai people, seen as a barbarous, homogeneous entity. The events depicted in the plot are impossible by their very nature, representing as they do the dystopian manifestation of a state of exception that does not spare the wealthy (see Hoad 2013). This state of exception appears as such only by virtue of a vision that does not consider the binary through which the existence of places *for* disaster becomes structural, as the 'exception' only comes into being when disaster affects those who belong to the place *of* disaster. The state of exception is not exceptional at all for those who are believed to belong to the place *for* disaster. Once again, it is when catastrophe trespasses into the safe space of civilisation that the place *for* disaster is defined as such, engendering the dystopian exception that could subvert and erase the world as the 'we' knows it. In the closing scene, Henry, Maria and their three children are on the plane that is bringing them back to the safe space of the 'master', wounded but alive, as if they were the worthy survivors of a catastrophe that was nothing but a process of natural selection to weed out the poor, the indigenous and the unfit.

Blindness, which I have discussed at length in three of my recent publications (Giuliani 2016a, 2017b, 2020), is about a sudden epidemic of blindness that starts in the city of São Paulo, Brazil. It belongs to the speculative fiction genre, since the story and its characters deal with events and phenomena resulting from the likely extrapolation of actual natural or scientific processes, actions and reactions, projected in a 'what if' dimension that as such may constitute a new framework for human deeds (Heinlein 1947, p. 19; Merril 2017, p. 22). Although blindness not only causes accidents and death but is also a metaphor for the 'eclipse of enlightenment' (Von Koppenfels 2004, p. 156),[16] it does not bring about the end of the world. Rather, it allows experiencing the end of the *world as the 'we' knows it* (e.g. driven by and organised according to the time and space of capitalism). In *Blindness*, the blind are forcefully removed from their homes and confined to an abandoned mental hospital where neither medical or psychological assistance nor basic necessities such as linens and soap are provided. The facility is patrolled by the army, who have received orders to kill anyone attempting to leave the grounds. Forced co-existence of people with different experiences of disaster triggers the abuse, rape and murder of the most vulnerable and unfamiliar with the 'state of exception'. A bad omen seems to emerge for humanity: that only the most violent, horrible and depraved beings will survive, and the post-apocalyptic world will be permanently split by a civil war – as in *The Road* (the novel as well as the film), where cannibalism is a metaphor for selfishness and lack of empathy. In

Blindness, when a group of men brutally rapes and kills a woman, the protagonist (the sighted wife of a doctor, played by Julianne Moore, who pretends to be blind) sets their ward on fire. They will all die. Once she discovers that the soldiers have become infected and left the facility, she leads the people from her ward, her new affective community, out of the hospital and into the chaotic city, where the dead and the living blind are left in a 'state of nature' characterised by a war of all against all for the limited resources available. Back at her and her blind husband's luxury apartment, the members of the small community enjoy a moment of peace and the morning after, the 'patient zero' of the epidemic regains his sight. This is a new beginning in their lives, marked by awareness that the vulnerable can only survive with the help of the stronger, healthier and wealthier.

The separation between the world of the wealthy and healthy and the world of the poor is at the core of Neill Blomkamp's dystopian film *Elysium*. The wealthy live on Elysium, a gigantic space habitat in Earth's orbit. The poor are trapped on polluted, overpopulated, desertified Earth, where they lack adequate medical care. Max Da Costa (Matt Damon) is an orphan who lives in a derelict neighbourhood of Los Angeles and works in the factory that makes Elysium's weaponry. After being exposed to radiation, he has only five days left and is determined to reach Elysium to get adequate care and survive. Earth's inhabitants are not allowed to enter Elysium. They are considered illegal immigrants and are routinely shot by the officers headed by Defense Secretary Delacourt (Jodie Foster) when their small spaceships approach the border – as happens on many other borders around the world. His personal struggle will become a struggle for equality and his sacrifice will allow a reboot of Elysium's computer and the registration of every Earth resident as an Elysian citizen.

Along the same lines, a successful Brazilian dystopian web television series, *3%* (created by Pedro Aguilera, broadcast since 2016, and distributed in 190 countries), deals with a spatially and socially divided society (the chosen few vs the poor and hopeless inhabitants of the slums called the Continent). In a 'survival of the fittest' kind of contest, every year 3 per cent of the poor who have reached the age of 20 are given a chance to access the world of the chosen few, provided they demonstrate superior mental and physical abilities. The *mundo do lado de lá* (the world on the other side called Maralto, the 'high sea') is inhabited by the ones who have been selected through intelligence and aptitude tests. The rewards make the candidates even more determined to pass the tests – real ideological weapons aimed at persuading them to leave their old life behind – and preserve the system. To prove they are 'deserving' and access the world of the fittest, they must be willing to renounce their fertility: the selection process cannot be affected by personal ties such as parenthood. This is the only way to preserve a sustainable world that is dominated by technoscience, highly patrolled and authoritarian, in which individuals are treated as gods and goddesses with no will nor desire other than those fulfilled by its institutions. Some of the candidates are rebels in disguise who belong to the Cause (the secret organisation conspiring against the Process, that is, the selection on which the separation of the worlds is based). Because of their plot to kill Process functionaries and subvert the order, members of the Cause are regarded as terrorists who, like the rebels in *Snowpiercer*, cannot understand that class division, along with the world shaped by the Process, is the only way to preserve humanity and cultivate its superiority. It is not a thrust for freedom, social justice or more generally well-being that drives the engine of revolt: the opportunity to move to the other world and leave misery behind is too good to miss for the sake of social equality. But the issue of infertility can undermine the candidates' faith in the Process: many lose their mind when they find out that they will have to leave their children behind and the final test before being granted access to the Maralto is a purification ritual that will prevent them from having 'a natural family' for the rest of their life.[17]

Fantasies of selection play out in the 2019 Australian–U.S. dystopian film *I Am Mother*, directed by Grant Sputore, in which robots grow human embryos to repopulate Earth after the extinction of humankind. Children are taught how to feel and act in accordance with the 'highest human values' – responsibility and solidarity. Fantasies of a new beginning are coupled in the film with the sinister idea that human genocidal instinct has reached its peak and is now embodied by a machine that turned against unfit humans. The solution to Anthropocenic individualism is a new Anthropocentric Renaissance. As in *3%*, survivors plotting against the army of robots are seen as perpetuating human wrongs and are left to die in a world that robots have deprived of food, shelter and peace. Here, too, unfit humans are exposed to famine, pollution, epidemics and violence in the name of inequality and a differential distribution of wealth, health and technoscience. Just as in Victorian India, where epidemics were spread to *exterminate* and *discipline* the population (Davis 2000), here too famine, pollution, epidemics, and violence produce a selection among humans aimed at reducing, if not eliminating altogether, the masses of the 'unfit' and building an eugenically engineered society of the 'elect'.

In the face of hungry hordes, global wars and terrorism over the control of fossil resources as well as the threat of planetary environmental catastrophe, fantasies of survival include isolating the wretched through new walled borders and other devices capable of re-establishing the distance. Protective boundaries are drawn to protect non-expendable subjects, guaranteeing them resources and a safe space where they can live untouched and avoid contamination, lest natural incidents like desertification occur in the places *of* disaster and the dystopian future become real. In speculative fiction, these boundaries are geographical and political as well as genetic. A good example is the dystopian film *Code 46* (2003) directed by Michael Winterbottom, where metropoles similar to Saskia Sassen's (1991, 1994) global cities are inhabited by the few who have travel documents. *Code 46* is a law that forbids sexual intercourse, marriage and procreation between individuals with genetic proximity – a likely occurrence, given that most of the global population is the result of cloning. In this scenario, the new beginning consists of a reassuring, utopian, eugenically engineered 'after' that pitches those 'worth saving' against the 'expendable', the drowned of the *lado de lá* for whom no pity is felt. Like in the case of Hurricane Katrina, in *The Impossible*, *Blindness* and *I Am Mother*, the official answer to catastrophe is, *homo homini lupus*. There is no society nor community taking care of its weakest members, nor any social obligations, only individuals and families that form part of a network of power relations projected onto a transnational and global level. In *Blindness*, these networks of relations generate affective communities capable of resisting the effects of the threat, creating solidarity and overcoming social differences. But this only lasts until the return to normalcy restores differentiation among people (*Blindness*). Trans-classist (but not transracial) emotional communities of survivors (*The Impossible*), as they might be termed, thus appear as mere means of survival and are inherently temporary, lasting only until the reunited upper-middle-class white family flies home to take back its rightful place in the Western world. This scenario is amplified in the case of a robot (*I Am Mother*) engineering a new humanity and keeping it secluded; when the 'deserving' youth are physically separated from their habitat and affects (*3%*); and when worthy humans rescued from the desertified 'out there' are spatially and genetically segregated and forced to live in artificial light (*Code 46*). Next, I will focus on ways of imagining the environmental apocalypse as 'living in thick times', to borrow from Haraway (2016, p. 2) – or living *through and with the catastrophe*.

7 *Trans-corporeal* and *intra-active* mutations

Very few films and popular narratives envision a world without humans. The Anthropocenic Anthropoi cannot imagine a world without them, feeling confident that they are the only life worth rescuing and surviving, and ascribing their superiority to their closer proximity to God or intellectual supremacy over other species. This sense of superiority comes from a conception of bodily and cognitive integrity that is either a reflection of God's perfection or the expression of normative ideas of body and mind. While human enhancement is acceptable because it responds to the needs of increased productivity and effectiveness, unruly metamorphoses have been generally regarded as *monstrous*. The need to remain untouched and un-contaminated is a moral, social and political imperative that since Modernity has taken different shapes and forms, as Foucault reminds us – hence mutilation inflicted by law, bodily modification due to illness, and racial miscegenation are seen as aberrations. As one of the most basic social values in Europe and its colonies, bodily integrity has shaped social engineering as much as medical science with the aim of preserving the best genetic traits, protecting the body and mind of the privileged from the dangers of the environment and the effects of value extraction. In fantasies of post-apocalyptic survival, the human body is still envisioned as untouched or at least not contaminated in its (genetic) essence. Below, I will challenge this imaginary through the British film *The Girl with All the Gifts* (2016), which narrates the survival of a metamorphosed human *through and with* the environmental apocalypse. Here mutation is the result of mutual contamination between the human body and the environment – the sublimation of a *trans-corporeality* (Alaimo 2008, 2010, 2018) that sutures *bios*, *zoe* and the inorganic (Esposito 2004; Braidotti 2013).

> Trans-corporeality, along with other theoretical concepts within feminist posthumanities, suggests a new figuration of the human after the Human, which is not founded on detachment, dualisms, hierarchies or exceptionalism, and which does not, in Val Plumwood's terms, 'background' nature (1993).
>
> (Alaimo 2018, p. 436)

Contrary to what the 'we' would expect, the protagonist does not fight mutation but rather takes it as evidence of the interdependency between humans and nature (Davis et al. 2019, p. 8) and as a permanent condition. In the film, directed by Colm McCarthy and based on Mike Carey's 2014 novel by the same name, humans infected by a mutated fungus, the *Ophiocordyceps unilateralis*, have turned into post-human cannibals called the 'hungries', who share many similarities with the fast-running rabid monsters I have described in Section 1. The virus eventually infects everything and everyone, Life and Nonlife. In a military base, some children are held captive and treated like *monsters*: One of them, Melanie (Sennia Nanua), guides the audience through her life at the base, day and night. The children are restrained in their cells, wear muzzles, and every morning are pushed on wheelchairs to the army base school where the teacher Helen Justineau (Gemma Arterton) is in charge of their education, or to the lab, where Dr Caroline Caldwell (Glenn Close) uses them for her experiments in hopes of finding a cure or a vaccine. These hybrid second-generation children, who crave living flesh but retain the ability to think and learn, are indeed humankind's only hope. Their nervous system has adapted to the fungus: they are *queers* whose 'nature' does not fit neatly into human classifications. When the base is attacked by a group of hungries, Melanie, Helen, Dr Caldwell, Sergeant Eddie Parks (Paddy Considine) and two soldiers get on an armoured truck and try to reach London. Melanie helps them find food and move around the city, where they find that a mass of mutated humans has turned into a giant plant that may release spores that would end humankind. After finding shelter in an abandoned mobile lab, one of the soldiers volunteers to go out and search for food, but he is ambushed by a mob of child hungries. To prevent the hungries from attacking 'her' humans,

Melanie acts as an animal and kills their leader. But Dr Caldwell, who is slowly dying of a blood infection, is still determined to open Melanie's head to observe the fungus and find a cure. Melanie runs away and sets the giant plant on fire in order for it to release its spores. In fact, Melanie has realised that she is not doomed to live in subordination to the humans. She and her hybrid feral companions *are* the future, a future in which there is no room for any human pretence of isolation and superiority over nature (Tsing 2015). Caldwell chases her but is killed by the child hungries. Parks leaves the lab, goes after Melanie and finds the giant plant on fire: he is quickly infected by the spores and asks Melanie to kill him. In the lab, Helen stands inside the sealed door, watching the spores fall. In the last scene, she is safe but imprisoned inside the sealed lab, from where she lectures all the child hungries who are now led by Melanie, teaching them how to become the heirs to humanity in what we may call a new post-Anthropocene era. However, although there is apparently no room left for value extraction and the *geontologies* that sustained it, Helen's teachings are based on the same principles that guided the Anthropocene: the future is open, and it is up to the spectator to guess what it will look like.

A parallel reading of *In the Flesh*, *Annihilation*, *Arrival*, and *The Girl* makes it clear that a world that does not retain at least some vestiges of humanity is unthinkable. In *In the Flesh*, relations with the Otherworld force humans to think about love, care, self-care, reciprocity and taking responsibility for the past and the present. In *Annihilation* and *Arrival*, endless and unpredictable mutations are metaphors for the encounter with the unknown Out there violating the dualism between catastrophe and disaster, human and less human, human and alien. Both films make an effort to imagine the encounter and its consequences as a 'change' to which we, the humans, need to resign (*Annihilation*) or as an 'exchange' that will increase human awareness (*Arrival*). While communication and mutation in the former are against humans' will, acceptance from wiser humans in the latter allows the whole community to benefit from an exchange that challenges all preconceived notions about humans' and the planet's life. As Lena says referring to the alien Shimmer, 'It wasn't destroying. It was changing everything. It was making something new'. In these films as in *The Girl*, the future is left unsaid and open to interpretation, including the possibility that the non-human environment may once again be taken over – materially and symbolically – by (post-)humans. Still, it is through the *human gaze* that the 'we' imagines and constructs the world after the end of the world. When Lena says that 'it *wasn't* destroying', she is implying that 'it' had been somehow defeated. Although her body bears the marks of genetic mutation, refraction and DNA recombination have apparently stopped outside the Shimmer. Similarly, the alien gift in *Arrival* might be seen – contrary to the reading I suggested in Section 2 – as a device for enhancing the human body and mind that does not challenge in any way the ontologies and logics of the Anthropocene.

So, if the imaginary of the environmental catastrophe cannot envision a world void of humans, what message are these films trying to convey? What does the experience of living *through and with* the catastrophe mean? On the one hand, these films seem to claim that, as Stacey Alaimo would put it, 'dwelling in the dissolve, where fundamental boundaries have begun to come undone, unravelled by unknown futures, can be a form of ethical engagement that emanates from both feminist and environmentalist practices' (2016, p. 2). They frame catastrophe into the (political) opportunity to change the very assumptions that support the ontologies and logics of the Anthropocene: the fact that humankind and the environment are inextricably linked becomes the very terrain of *the political*. Louise, Lena and Melanie are *trans-corporeal* and *intra-active* embodiments of Marx's argument that '[t]hat man's physical and spiritual life is linked to nature means simply that nature is linked to itself, for man is a part of nature' (1844, p. 41). In fact, Melanie's subjectivity is not only *in a trans-corporeal relation* with the fungus but is also preceded by biological *intra-active* contamination of the human by the fungus.

These films' political stance for *trans-corporeality* and *intra-activity*, their taking criticism of the Anthropocene to the extreme – condemning extermination of *risky bodies* and arguing in favour of the preservation of *bodies at risk* (Giuliani 2016b; Salerno 2016), whether hybrid or alien – may be regarded not only as a new take on the relation between care, self-care and earth-care but also as a call for the end of the Anthropocene. But while Louise's effort in *Arrival* to communicate with (and care for) the alien is motivated by self-care – that is, the desire to save the world – self-preservation in *The Girl* means destroying the world to save the planet: Melanie's actions trigger human extinction and lead to an unspeakable future.[18] What a future without humanity means for her and her fellow hungries is the opportunity to live freely as hybrids and build a trans-speciesist *kin* with the *monstrous* fungus. The end of humankind inaugurates a post-human epoch based on the *awareness* that life is inherently contaminated and interdependent. A creature of the Chthulucene, Melanie is that 'material body [that] always resides within some *assemblage* or other, and its thing-power is a function of that grouping' (Deleuze and Guattari 1980). She is a thing that 'has power by virtue of its operating in conjunction with other things' (Bennett 2004, pp. 353–354). In the words of Anna Tsing, the catastrophe that produces Melanie and the hybrids is offered as a dystopian *revelation* of the fact that:

> We are contaminated by our encounters; they change who we are as we make way for others. As contamination changes world-making projects, mutual worlds – and new directions – may emerge. Everyone carries a history of contamination; purity is not an option. One value of keeping precarity in mind is that it makes us remember that changing with circumstances is the stuff of survival.
>
> (Tsing 2015, p. 27)

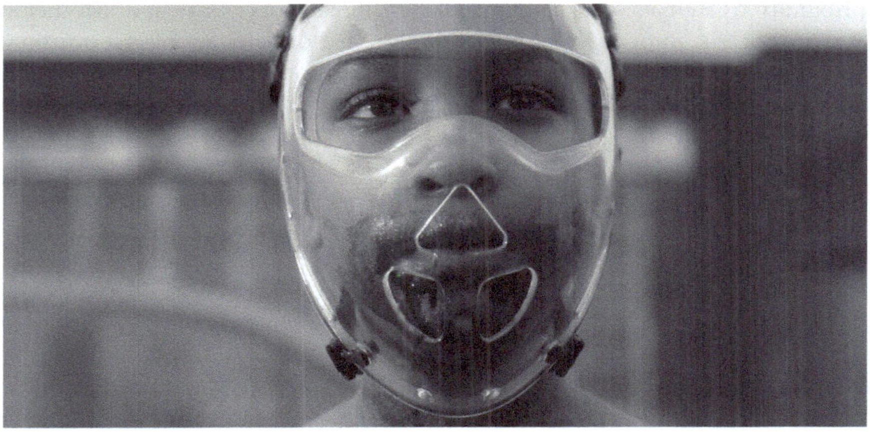

Figure 3.3 Still from *The Girl with All the Gifts* (2016), directed by Colm McCarthy, showing the protagonist, Melanie (Sennia Nenua).

If we compare this scenario to the one envisioned in *I Am Mother*, we see how in the latter the catastrophe triggered by the robots is meant to prevent unruly modifications and produce 'purity'. Earth-care and self-care are denied, while care is restricted to the 'pure and superior humans'. In the world that 'Mother' has designed for humanity there is no room for Jean Luc Nancy's idea of *exposure* to the Other (1996, p. 18), because the Other – the unfit – is left out and will be exterminated to safeguard the supremacy and pristine life of the enhanced. Nor can there be an *exposure* of human flesh that leaves room for 'multispecies liveliness to disperse and displace human exceptionalism' (Alaimo 2016, p. 1). In this film, intensive farming by robot soldiers seems to indicate that a new and superior humanity will necessarily reproduce the Anthropocenic rule – both over its technological expression (the robots) and over the entire planet. In line with patriarchal normativity, self-care is deployed here as a Foucauldian 'technology of power', that is, as a means of oppression, containment, control and subjection imposed by the hyper-technological Anthropocenic human-like robot 'Mother' on her human daughter. The loving and caring 'Mother' sets limits on the little girl's autonomy and independence. 'Daughter' does not need to know what is happening outside the base and cannot have an opinion on the robots' actions. Her body is not hers but a device that is meant to guarantee her *perfection* according to standards imposed by 'Mother'. She cannot experience self-love, sexuality and pleasure unless within those boundaries: she will not be able to 'luxuriate' – in Haraway's parlance – nor reproduce with other humans, because all the embryos from which the 'new' children are generated are genetically contiguous. Her only possibility to feel and love is within parameters established by 'Mother'.

Post-human Melanie, on the contrary, becomes *feral*, and her closer proximity to animality signals the denial of human superiority and 'Mother's' Anthropocenic imperative of purity and perfection. Her act of insubordination seems to imply that the end of the Anthropocene may mark the beginning of a new era. This is what it means for her to live *through and with the catastrophe*:

> [S]taying with the trouble requires learning to be truly present, not as a vanishing pivot between awful or edenic pasts and apocalyptic or salvific futures, but as mortal critters entwined in myriad unfinished configurations of places, times, matters, meanings.
>
> (Haraway 2016, p. 2)

In line with Haraway's perspective (see also 2015), Melanie's metamorphosis does not necessarily frame the human past as either awful or Edenic: her kind retains both the memory of the *monstrous* history of Anthropocenic horrors, epitomised by the fungal mutation and the forced confinement of the hybrid children, and the human *ability to learn and be kind* – clearly evident in Justineau's behaviour. In the case of Melanie and her fellow hybrid post-humans, these abilities come with the *human features* they have retained. And yet, as Barca has suggested, the war that the hybrids wage on humans assumes the existence of only two worlds: the world of the 'we', represented by Western military, and that of the *monstrous* hybrids. Haraway's argument in the quote above recalls Escobar's *pluriverse* as well as the different temporalities and modernities discarded by the dominant order, or One World World's Modernity (Fabian 1983; Dussel 1995; Mignolo and Walsh 2018; Swyngedouw and Ernstson 2018). In Melanie's post-apocalyptic world there is no room for them. The last scene, in which Justineau, the only human allowed to survive, is teaching the feral hybrids while Melanie is forcing them to listen and learn, can be read through a postcolonial gaze as replicating the colonial paradigm, with the fittest teaching *barbarians how to achieve Anthropocenic humanness through knowledge of the civilised world* – or it may be seen as the moment when the colonial subaltern learns from their master how to rule the word by the same means.

8 Conclusions

In this section, I have focused my analysis on understanding catastrophe as the moment in which the *monstrosity* of the Anthropocene is *revealed*. This *revelation* is brought in by another *revelation*, that the 'we's' fears of catastrophe are indeed fears of losing the violent Anthropocenic discursive *dispositifs* that made its space and time safe. These *dispositifs* correspond to classic dualisms of culture/nature, human/animal, human/non-human, organic/inorganic, Here/Out there, civilised/barbarian, centre/peripheries that have structured an hegemonic conception of the world since colonial Modernity, and especially since ideas of *human progress*, *bodily enhancement* and *social renewal* have begun shaping Anthropocenic bio- and necropolitics. The idea that borders, walls and segregated spaces built on and reproducing these dualisms cannot stop a catastrophe, and that 'barbarity' will be freed from its constraints, is unconceivable, and that is the reason why there seems to be no alternative to the Anthropocene. Rather than a public reflection on the horrors of the Anthropocene and its own monstrous barbarism, what is at the core of scientific, political and cultural discourses of environmental catastrophe is the monstrification of places *for* disaster (that is, where the polluting effects of the Capitalocene are most devastating) and the peoples who inhabit them, deemed unable to master the crisis.

Through an analysis of the meaning attached to natural catastrophes in Western culture, I have explored the shift from a perception of catastrophe as God's punishment to one of *renewal* in the *siècle de Lumières*. Although the imaginary of environmental catastrophe does not envision the possibility that the ultimate crisis may be overcome, post-apocalyptic scenarios in scientific as well as in popular culture still bear the imprint of the Anthropocenic Anthropos.

A post-apocalyptic future where modern distinctions and dualisms have been erased is unthinkable (and hence the future is envisioned as humanless) not only because the prospect of broken promises and shattered hopes of boundless improvement and unlimited natural resources is dismissed as unacceptable, but also because the 'we' is unwilling to come to terms with the wrongs of the Anthropocene, least of all the violence imposed on the *less-human*, the *other-than-human* and *inorganic life*. What the 'we' is unwilling to acknowledge is that the racialised, the gendered and the poor have been violently constructed as *less-human* and in need of white paternal guidance, and hence deprived of recognition as self-governing subjects; *other-than-human* plant and animal species are seen merely as a resource to exploit for the benefit of human life; and Earth has been reduced to a fossil reservoir of *inorganic life*.

As the looming threat of vanishing *history* and *geography*, signalled by mass migrations and terrorism, hovers over the safe space of the 'we', the only relief is found in the idea that the environmental catastrophe and its effects on the world could be limited in space and time. In all the films I have analysed, a safe space is always provided where the 'we' can find shelter and be reassured about its privileges and certainties. Some (*The Core* and *The Wandering Earth*) celebrate human victory over the catastrophe with the help of techno-scientific knowledge and weapons. Others (*The Day after Tomorrow* and *Snowpiercer*) emphasise a resilient approach and success in saving as many *lives* as possible or resisting the urge to succumb to the imperatives of *homo homini lupus* in a context devoid of institutional guide, social norms and moral obligations (*Blindness*). None of them question the bio- and necropolitics at the core of the Anthropocene and its history of horrors, but instead reproduce its mechanisms: dystopian versions of Malthusianism and social Darwinism are seen by most as the only way out of the crisis and are used, as they were in the past, as means of *renewal*. The dystopian society in *3%*, set in the firm belief that the dual social system will help preserve the *lado de lá* from the polluted world Out there, resorts to administering aptitude tests to select and preserve humanity's 'best specimens'. Fantasies of selection *and* colonisation, of new plantations *and* new forms of slavery take centre stage in many of these films and TV series. They are also at the core of *Elysium*'s social critique: the masses are left behind on desertified, polluted, violence-ridden Earth, while the chosen few live on a gigantic space habitat. Two films, *Code 46* and *I Am Mother*, do not shy away from describing eugenics as a means to the survival of social hierarchies based on the well-being of the fittest, and necropolitics as a way of governing the expendable. The expendable, the poor and most vulnerable members of society described by the segregated elite as miserable criminals, are left Outside the safe spaces of global cities and robot bases. A critique of growth and development, with their logics of exploitation and extractivism, is absent in most films: all hopes of a new beginning for human society seem to be based on the old principles of value extraction and inequality. In European and Western popular and hegemonic imaginary of catastrophe, the idea of *renewal* does not go much farther than making a blanket accusation against mankind as a whole – as if humans were all responsible to the same degree – and suggesting minor corrections to the capitalist model. In these films as much as in the normalised discourse of climate change, the *monstrosity* of the Anthropocene emerges *unquestioned* in its enduring *violence* against everything that is considered unworthy and expendable. As Laura Pulido notes, referring to the outcome of the 2009 Copenhagen Climate Change Conference and the decision of limiting the maximum global average temperature increase to no more than 2 degrees Celsius above pre-industrial levels, 'this moment illuminated the racial geography of global warming, our lack of political will, our disregard for nonwhite and poor lives, and the deeply immoral nature of the Anthropocene' (2018, p. 120).

What I tried to do here is add a postcolonial perspective to the existing dialogue between political philosophy, cultural studies and environmental humanities, especially as regards the current debate surrounding degrowth, post-development and social reproduction. More emphasis is needed on crucial issues of care, earth-care and self-care for an adequate political critique of the Anthropocene (Barca 2020). I argued for the necessity of claiming *trans-corporeality*, *intra-activity* and interdependency of Life and Nonlife as the new fundaments of *the political*. The politicisation of *trans-corporeality* and *intra-activity* – in line with ecology movements and indigenous cosmologies and forms of resistance – is necessarily based on a reflection on the *fundamental political value* of the interdependency between care, self-care and earth-care. In order to rethink the political struggle against capitalism and the Anthropocene as a feminist struggle, the *monstrosity* of patriarchal reproduction of care and self-care as technologies of power needs to be unpacked. Only through the mutual limitation of care and self-care and their re-framing as earth-care will the *monstrosity* of the Anthropocene *reveal* itself, giving way to a new temporality made of multiple non-extractivist epistemologies and *pluriverse*. To better explain my idea of *trans-corporeality* and *intra-activity* as political values, I offered a reading of the British apocalyptic film *The Girl with All the Gifts*, where the catastrophe is caused by a genetically modified fungus. Humans undergo a metamorphosis that leads to their death or to permanent mutation into post-human hybrids whose brain is enveloped by the fungal symbiont. Through the figure of a little girl, Melanie, and her fellow hybrids, I discussed the impossible distinction between care, earth-care and self-care in the context of the vanishing ideal of 'pristine and pure' humanity. Because of the symbiosis between the human body and the vegetal organism, the fungus cannot be eradicated without harming the hybrids and vice-versa. Thus self-care as self-protection means mutual care among the hybrids as well as earth-care towards non-human organisms. Self-care, care and earth-care are preserved against the 'we' in a war to the death that prevents the return of the Anthropocene. *Here, the monstrous Anthropocene emerges again in all its evilness*: as the children of the Chthulucene, Melanie and her fellow little hungries are seen by humans as *monsters* whose only purpose is to be studied, sectioned and killed for the sake of humankind's survival. The world that Melanie envisages, or that I see as envisaged in the ultimate gesture of the little girl igniting the tree and liberating its spores, is instead one that allows understanding the *generativity* of *intra-activity* 'as a web of biological, social and cultural dimensions' (Bauhardt 2019, p. 29). Nonetheless, this new world entails both a caesura and the promise of a new One World: Melanie and her fellow feral hybrids create a new society in which no *pluriverse* is allowed nor a critique of the knowledge sustaining the Anthropocene is envisaged. On the contrary, the last scene, which sees the last human teacher lecturing the feral hybrids, recalls old 'scenes of subjection' of colonised 'barbarians' learning from and against their oppressors how to master the planet.

Notes

1 Introducing her exhibition *Big Fence/Pitcairn Island* (The Photographers' Gallery, London, 2019), on Britain's last overseas territory in the Pacific, inhabited by the descendants of the Bounty's crew, Irish artist Rhiannon Adam wrote: 'Despite the reality of its imposing cliffs and lack of beach, to many outside observers, Pitcairn epitomises Utopia – a vision cultivated by Hollywood's romanticised adaptations of the "Bounty" story. In 2004, this façade slipped, when a series of child sexual abuse allegations emerged, leading to convictions of eight Pitcairn men, including the current mayor. Despite best

efforts to repopulate, by 2015, fewer than forty islanders and just one child remained'. https://thephotographersgallery.org.uk/whats-on/exhibition/tpg-new-talent.

2 'l'hypotèse d'une nature rationnalisée, maîtrisée (certains diront: domestiquée) et brusquement livrée à une maîtrise scientifique et technique qui en épuise les ressources'.

3 'Privilégie [...] l'idée d'une destruction catastrophique qui mènerait la societé humaine à se sauver par l'intermède de la pensée raisonnable at du progress scientifique'.

4 'que la Nature et le lois physiques de l'univers pourront être responsables d'un éventuel obstacle au progrès'.

5 'les progrès de cette perfectibilité [sont] désormais indépendants de toute puissance qui voudrait les arrêter'.

6 'Se impliquer dans l'histoire'.

7 www.floornature.com/creation-from-catastrophe-exhibition-at-the-architecture-gallery-london-11360/.

8 'Kant observe qu'il valait mieux, dans les endroits sujets à des tremblements de terre, bâtir moins haute, et se server de materiaux plus légers'. In general, see *History and Physiography of the Most Remarkable Cases of the Earthquake which Towards the End of the Year 1755 Shook a Great Part of the Earth* (1756) and *Critique of Practical Reason* (1788).

9 The list is impressive and includes the typhoon Haiyan in the Philippines (2013), which caused 10,000 victims and involved 4 million people, the earthquake in Nepal (2015), which caused 11,500 victims and 3.5 million displaced people, and the seaquake in Sulawesi (2018), which caused 4,500 victims and over 200,000 displaced people.

10 See www.bhopal.net/.

11 The event and its consequences inspired Bradley Parker's film *Chernobyl Diaries* (2012) – based on the novel *The Diary of Lawson Oxford* by Israeli writer Oren Peli, who also directed the very successful horror film *Paranormal Activity* (2007).

12 See, for instance, Annie Sprinkle and Beth Stevens' *sexecological marriages*, one of which, the blue-themed, I attended at the Venice Biennale in 2009. 'The wedding series is part of Sprinkle and Stephens's exploration of "sexecology," an art and activist project that presents the earth as lover, partner in sexual adventures, source and receiver of polymorphous pleasures. Comprising documentary films, performances, walking tours, manifestoes, and academic research, Sprinkle and Stephens's multimedia artwork celebrates pollen-amorous practices ranging from tree-hugging to "grassilingus"'. (Tola 2019, p. 231). See also Stephens and Sprinkle (2011) and Stephen, Sprinkle and Morris (2016).

13 For insights into the international debate on eugenics, see the works of Claudio Pogliano (1984; 2005). For a genealogy of eugenics in the United States, see the referenced works of Edwin Black, Jonathan Peter Spiro (2008), and Wendy Kline (2005). On eugenics in the British Empire, Australia and the United States, see the works of Diane Wyndham (2003), Nancy Stepan (1982), Philippa Levine and Alison Bashford (2010); on eugenics in Italy, see Claudio Pogliano and Francesco Cassata (2006); and in Sweden, Luca Dotti (2005).

14 See Spike Lee's documentary on Katrina produced by HBO, *When the Levees Broke. Requiem in Four Act* (2006); the last chapter in Kevin Rozario's *Culture of Disaster*; and the debate on the website 'Understanding Katrina. Perspectives from the social sciences' (http://understandingkatrina.ssrc.org).

15 http://understandingkatrina.ssrc.org/Gilman/.

16 Von Koppenfels refers here to José Saramago's novel, written after the writer had witnessed the early 1990s civil war in the Balkans.

17 A remarkable British–U.S. film has thematised the issue of infertility and connected it to migration and global inequality. *The Children of Men* (2006), directed by Alfonso Cuarón, is set in 2027 when after 18 years of global human infertility, a refugee woman, Kee, discovers she is pregnant. Theo Faron (Clive Owen) will help her avoid being executed for being an illegal migrant or kidnapped for fertility experiments.

18 In a 'war of the worlds' meets *mors tua vita mea* kind of scenario, Melanie brings the lives of the Anthropocenic humans to an end, somehow providing a dystopian version of Haraway's claim about sustainable population growth, that is, a *responsible* relationship between organic beings and inorganic formations to challenge patriarchy, capitalist exploitation and extractivism (2016, p. 2). Like other films analysed in this section, *The Girl* advocates a dramatic reduction of the human population on Earth. Nonetheless, birth control and segregation (*3%* and *Snowpiercer*), genetic manipulation of human embryos (*Code 46* and *I Am Mother*) and 'natural selection' via catastrophe are very problematic aspects of the Anthropocenic mentality based on a patriarchal, classist and racist vision of the after-Armageddon (see Broderick 1993). Furthermore, this vision is grounded in a memory of the future based on the specific history of peoples and groups who experienced reproductive coercion (e.g. slave breeding and birth-control programs targeting African Americans or citizens of Asian and African descent in postcolonial white societies). Therefore, post-apocalyptic dystopian narratives like *The Girl* could be the object of criticism for triggering traumatic memories of procreation as a means of subjection and coupling with 'the sterilising narrative of wiping the world clean by apocalypse or salvation'. Nevertheless, the richest humus for inquiries into our present remains, as Donna Haraway reminds us, 'science fiction and fantasy, speculative fabulation, speculative feminism, and string figures. Blocking the foreclosures of utopias, [they] kept politics alive' (2016, p. 150).

References

Abbott, P. (2006) 'Chernobyl: living with risk and uncertainty', *Health, Risk & Society* 8 (2), pp. 105–121.

Agostini, L. (1580) *La Repubblica Immaginaria*. Turin: Ramella (this edition 1580).

Ahmed, S. (2000) *Strange Encounters*. Abingdon and New York: Routledge.

Ahmed, S. (2014) 'Selfcare as warfare'. *Feminist Killjoy*. [Blog] Available at: https://feministkilljoys.com/2014/08/25/selfcare-as-warfare/ Accessed 11th March 2020.

Alaimo, S. (2008) 'Trans-corporeal feminism and the ethical space of nature'. In: Alaimo, S. and Hekman, S. (eds) *Material Feminisms*. Bloomington: Indiana University Press, pp. 237–264.

Alaimo, S. (2010) *Bodily Natures: Science, Environment, and the Material Self*. Bloomington: Indiana University Press.

Alaimo, S. (2016) *Exposed. Environmental Politics and Pleasures in Posthuman Times*. Minneapolis: Minnesota University Press. http://doi.org/10.5749/minnesota/9780816621958.001.0001

Alaimo, S. (2018) 'Trans-corporeality'. In: Braidotti, R. and Hlavajova, M. (eds) *Posthuman Glossary*. New York and London: Bloomsbury, pp. 435–438. http://doi.org/10.5040/9781350030275

Armiero, M. (2011) *A Rugged Nation: Mountains and the Making of Modern Italy*. Cambridge, MA: White Horse Press. https://doi.org/10.1659/mrd.mm096

Armiero, M. and de Rosa, S. (2016) 'Political effluvia. Smells, revelations, and the politicization of the daily experiences in Naples, Italy'. In: Thorpe, J., Rutherford,

S. and Sandberg, L. A. (eds) *Methodological Challenges in Nature-Culture and Environmental History Research*. London and New York: Routledge, pp. 173–186.

Armiero, M. and Fava, A. (2016) 'Of human, sheep and dioxin. A history of contamination and transformation in Acerra, Italy'. *Capitalism, Nature and Socialism* 27 (2), pp. 67–82. https://doi.org/10.1080/10455752.2016.1172812

Armiero, M. and Turhan, E. (2019) 'Le frontiere dell'Antropocene'. *Jacobin Italia* 4, pp. 36–39.

Arruzza, C., Bhattacharya, T. and Fraser, N. (2019) *Feminism for the 99 Percent: A Manifesto*. London: Verso.

Asad, T. (2007) *On Suicide Bombing*. New York: Columbia University Press.

Bacon, F. (1627) *The New Atlantis*. Cambridge, UK: The University Press (this edition 1909).

Baldwin, A. (2017) 'Rearranging desire. On whiteness and heteronormativity'. In: Baldwin, A. and Bettini, G. (eds) *Life Adrift: Climate Change, Migration, Critique*. London and New York: Rowman & Littlefield International, pp. 211–226.

Baldwin, A. and Bettini, G. (2017) 'Introduction: life adrift'. In: Baldwin, A. and Bettini, G. (eds) *Life Adrift: Climate Change, Migration, Critique*. London and New York: Rowman & Littlefield International, pp. 1–22.

Barad, K. (2007) *Meeting the Universe Halfway: Quantum Physics and the Entanglement of Matter and Meaning*. Durham: Duke University Press. http://doi.org/10.1215/9780822388128

Barad, K. (2008) 'Posthumanist performativity: toward an understanding of how matter comes to matter'. In: Alaimo, S. and Hekman, S. (eds.) *Material Feminisms*. Indianapolis: Indiana University Press, pp. 120–156.

Barca, S. (2011) 'Energy, property, and the industrial revolution narrative'. *Ecological Economics* 70 (7), pp. 1309–1315. https://doi.org/10.1016/j.ecolecon.2010.03.012

Barca, S. (2017) 'The labor(s) of degrowth'. *Entitle*. [Blog] Available at: https://entitleblog.org/2017/01/31/the-labors-of-degrowth/ Accessed 13 April 2020.

Barca, S. (2020 forthcoming) *Forces of Reproduction. Notes for a counter-hegemonic Anthropocene*. Cambridge, UK: Cambridge University Press.

Barca, S. and Leonardi, E. (2016) 'Working-class communities and ecology. Reframing environmental justice around the Ilva steel plant in Taranto, Apulia (Italy)'. In: Shaw, M. and Mayo, M. (eds) *Class and Community Development*. Bristol and Chicago: Policy Press/University of Chicago Press, pp. 59–76.

Bauhardt, C. (2019) 'Nature, care and gender: feminist dilemmas'. In: Bauhardt, C. and Harcourt, W. (eds) *Feminist Political Ecology and the Economics of Care: In Search of Economic Alternatives*. London: Routledge, pp. 16–35.

Benjamin, W. (1940) 'Theses on the philosophy of history'. In: Benjamin, W. *Illuminations: Essays and Reflections*. New York: Schocken Books, pp. 253–264 (this edition 1968).

Bennett, J. (2004) 'The force of things: steps toward an ecology of matter'. *Political Theory* 32 (3), pp. 347–372. https://doi.org/10.1177/0090591703260853

Benyus, J. (1997) *Biomimicry: Innovation Inspired by Nature*. New York: Harper & Collins.

Bettini, G. (2013) 'Climate barbarians at the gate? A critique of apocalyptic narratives on "climate refugees"'. *Geoforum* 45, pp. 63–72. https://doi.org/10.1016/j.geoforum.2012.09.009

Bettini, G. (2020 forthcoming) 'The end(s) of subjectivity and the abyss of the future in (climate) dystopias'. In: Giuliani, G. (ed.) 'Cidadania em perigo: crime, fim-do-mundo e biopolítica nas literaturas e no cinema pós-coloniais'. *ecadernos* 32.

Bhambra, G. K. (2007) *Rethinking Modernity: Postcolonialism and the Sociological Imagination*. London: Palgrave. https://doi.org/10.1057/9780230206410

Black, E. (2012) *War Against the Weak: Eugenics and America's Campaign to Create a Master Race, Expanded Edition*. Washington: Dialog Press.

Borghi, L. (2014) 'Assemblaggi affettivi. L'amore al tempo del quantoqueer'. In: Giuliani, G., Galetto, M. and Martucci, C. (eds) *L'amore al tempo dello tsunami. Affetti, sessualità, modelli di genere in mutamento*. Verona: Ombre Corte, pp. 207–216.

Boulding, K. E. (1966) 'The economics of the coming spaceship Earth'. In: Jarrett, H. E. (ed.) *Environmental Quality in a Growing Economy*. Baltimore, MD: Resources for the Future/Johns Hopkins University Press, pp. 3–14.

Braidotti, R. (2013) *Post-human*. Cambridge, UK: Polity Press

Broderick, M. (1993) 'Surviving Armageddon: beyond the imagination of disaster'. *Science fiction* 61 (20), pp. 342–382.

Butler, J. (2009) *Frames of War. When Is Life Grievable?* New York: Verso.

Byrd, J. (2011) *Transit of Empire: Indigenous Critiques of Colonialism*. Minneapolis: University of Minnesota Press.

Campanella, T. (1623) *The City of the Sun*. London: Journeyman (this edition 1981).

Canetti, E. (1960) *Crowds and Power*. New York: The Viking Press (this edition 1962).

Carey, M. (2014) *The Girl with All the Gifts*. London: Orbit Books.

Casid, J. H. (2018) 'Necropolitics at sea'. In: Gualtieri, C. (ed.) *Migration and the Contemporary Mediterranean. Shifting Cultures in Twenty-First Century Italy and Beyond*. Oxford, Bern, Berlin, Brussels, New York and Wien: Peter Lang, pp. 193–214.

Cassata, F. (2006) *Molti, sani e forti. L'eugenetica in Italia*. Turin: Bollati Boringhieri.

Centemeri, L. (2006) *Ritorno a Seveso. Il danno ambientale, il suo riconoscimento, la sua riparazione*. Milan: Bruno Mondadori.

Clark, N. (2011) *Inhuman Nature: Sociable Life on a Dynamic Planet*. London: Sage.

Córdoba Vargas, C. A., Romero, S. H. and Sicard, T. L. (2019) 'Key points of resilience to climate change: a necessary debate from agroecological systems'. *Climate and Development* (9), pp. 1–11. https://doi.org/10.1080/17565529.2019.1664376

Crosby, A. W. (1986) *Ecological Imperialism. The Biological Expansion of Europe, 900–1900*. Cambridge, UK: Cambridge University Press (this edition 2004).

Danowski, D. and Viveiros de Castro, E. (2017) *The Ends of the World*. Cambridge, UK: Polity Press. https://doi.org/10.4324/9780203007723

Davis, H. and Todd, Z. (2017) 'On the importance of a date, or decolonizing the Anthropocene'. *ACME: An International Journal for Critical Geographies* 16 (4), pp. 761–780.

Davis, J., Moulton, A. A., Van Sant, L. and Williams, B. (2019) 'Anthropocene, capitalocene, … plantationocene?: a manifesto for ecological justice in an age of global crises'. *Geography Compass* 13 (5), e12438. https://doi.org/10.1111/gec3.12438

Davis, M. (1990) *City of Quartz: Excavating the Future in Los Angeles*. London and New York: Verso.

Davis, M. (1998) *Ecology of Fear: Los Angeles and the Imagination of Disaster*. New York: New Press.

Davis, M. (2000) *Late Victorian Holocausts. El Niño Famines and the Making of the Third World*. New York: Verso.

Davis, M. (2002) *Dead Cities, And Other Tales*. New York: New Press.

Davis, M. (2005) *The Monster at Our Door: The Global Threat of Avian Flu*. New York: New Press.

De Condorcet, N. (1795) *Esquisse d'une tableaux historique deas progrès de l'esprit humain*. Paris: Flammarion (this edition 1988).

De la Cadena, M. (2015) 'Uncommoning nature'. *E-Flux Journal* 65, pp. 1–8. https://doi.org/10.11606

Deleuze, G. and Guattari, F. (1980) *A Thousand Plateaus: Capitalism and Schizophrenia*. Minneapolis: University of Minnesota Press (this edition 1987).

Diamond, J. (1997) *Guns, Germs, and Steel: The Fates of Human Societies*. New York: W.W. Norton & Co Inc.

Dicks, H. (2016) 'The philosophy of biomimicry'. *Philosophy and Technology* 29 (3), pp. 223–243. https://doi.org/10.1007/s13347-015-0210-2

Diderot, D. (1770) *Contributions à l'Historie de deux Indes*. Siena: Studi sull'Illuminismo (this edition 1976–1977).

Diderot, D. (1780–1782) *Le voyage en Hollande*. Paris: Maspéro (this edition 1982).

Dotti, L. (2005) *L'utopia eugenetica del welfare state svedese (1934–1975). Il programma socialdemocratico di sterilizzazione, aborto e castrazione*. Soveria Mannelli: Rubbettino.

Dubois, L. and Casimir, J. (2010) 'Reckoning in Haiti'. Available at: http://forums.ssrc.org/haiti/author/jean-casimir/

Dussel, E. (1995) *The Invention of the Americas. Eclipse of 'the Other' and the Myth of Modernity*. New York: Continuum.

Dynes, R. R. (2000) 'The dialogue between Voltaire and Rousseau on the Lisbon earthquake: the emergence of a social science view'. *International Journal of Mass Emergencies and Disaster* 28 (1), pp. 97–115.

Escobar, A. (2014) *Sentipensar con la Tierra: Postdesarrollo y Diferencia Radical*. Medellín: Universidad Autónoma Latinoamericana.

Escobar, A. (2016) 'Thinking-feeling with the Earth: territorial struggles and the ontological dimension of the epistemologies of the south'. *Revista de Antropología Iberoamericana* 11 (1), pp. 11–32. http://doi.org/10.11156/aibr.110102e

Esposito, R. (2004) *Bíos: Biopolitics and Philosophy*. Minneapolis: University of Minnesota Press (this edition 2008).

Fabian, J. (1983) *Time and the Other. How Anthropology Makes Its Object*. New York: Columbia University Press.

Federici, S. (2004) *Caliban and the Witch: Women, the Body and Primitive Accumulation*. Brooklyn, NY: Autonomedia.

Fornet-Betancourt, R., Becker, B., Gomez-Müller, A. and Gauthier, J. D. (1987) 'The ethic of care for the self as a practice of freedom: an interview with Michel Foucault on January 10, 1984'. *Philosophy Social Criticism* 12, pp. 112–131. https://doi.org/10.1177/019145378701200202

Foucault, M. (1961) *History of Madness*. London and New York: Routledge (this edition 2006).

Foucault, M. (1976) *Society Must Be Defended*. New York: Picador (this edition 2003).

Foucault, M. (1976) *The History of Sexuality: 1: The Will to Knowledge*. New York: Penguin (this edition 1998).

Foucault, M. (1984) *The Courage of the Truth (The Government of Self and Others II): Lectures at the Collège de France, 1983–1984*. New York: Picador (this edition 2008).

Fraser, N. (2014) 'Behind Marx's hidden abode'. *New Left Review* 86, pp. 55–72.

Friedlander, H. (1997) *The Origins of Nazi Genocide. From Euthanasia to the Final Solution*. Chapel Hill: University of North Carolina Press.

Giuliani, G. (2016a) *Zombie alieni e mutanti. Le Paure dall'11 settembre a oggi*. Florence–Milan: Le Monnier/Mondadori Education.

Giuliani, G. (2016b) 'Monstrosity, abjection, and Europe in the War on Terror'. *Capitalism Nature Socialism* 27 (4), pp. 96–114. http://doi.org/10.1080/10455752.2016.1192212

Giuliani, G. (2017a) 'The colour(s) of Lampedusa'. In: Proglio, G. and Odasso, L. (eds) *Border Lampedusa. Subjectivity, Visibility and Memory in Stories of Sea and Land*. Cham: Palgrave Macmillan, pp. 67–85. http://doi.org/10.1007/978-3-319-59330-2_5

Giuliani, G. (2017b) 'Afterword: life adrift in a postcolonial world'. In: Baldwin, A. and Bettini, G. (eds) *Life Adrift: Climate Change, Migration, Critique*. London and New York: Rowman & Littlefield International, pp. 227–242.

Giuliani, G. (2020 forthcoming) 'Monstrous beauties: bodies in motion between colonial archives and the migrant and refugee crisis'. In: Tate, S. A. (ed.) *The Palgrave Handbook on Critical Race and Gender*. London: Palgrave.

Goodman, J. and Salleh, A. (2013) 'The "green economy": class hegemony and counter-hegemony'. *Globalizations* 10 (3), pp. 411–424. https://doi.org/10.1080/14747731.2013.787770

Gorz, A. (2010) 'The exit from capitalism has already begun'. *Cultural Politics* 6 (1), pp. 5–14. http://doi.org/10.2752/175174310X12549254318629

Grosz, E. (2011) *Becoming Undone: Darwinian Reflections on Life, Politics and Art*. Durham, NC: Duke University Press.

Haraway, D. J. (2015) 'Anthropocene, Capitalocene, Plantationocene, Chthulucene: making kin'. *Environmental Humanities* 6, pp. 159–165. http://doi.org/10.1215/22011919-3615934

Haraway, D. J. (2016) *Staying with the Trouble. Making Kin in the Chthulucene*. Durham: Duke University Press. https://doi.org/10.1215/9780822373780

Haraway, D. J., Ishikawa, N., Gilbert, S. F., Olwig, K., Tsing, A. L. and Bubandt, N. (2015) 'Anthropologists are talking about the Anthropocene'. *Ethnos* 81 (3), pp. 535–564. https://doi.org/10.1080/00141844.2015.1105838

Hardin, G. (1968) 'The tragedy of the commons'. *Science, New Series* 162 (3859), pp. 1243–1248. http://doi.org/10.1126/science.162.3859.1243

Harrington, J. (1656) *Oceana*. In: Harrington, J. *The Commonwealth of Oceana and A System of Politics*. Cambridge, UK: Cambridge University Press (this edition 1992).

Hartman, S. V. (1997) *Scenes of Subjection: Terror, Slavery, and Self-Making in Nineteenth-Century America*. Oxford: Oxford University Press.

Harvey, D. (1974) 'Population, resources and the ideology of science'. *Economic Geography* 50, pp. 256–277. http://doi.org/10.2307/142863

Heinlein, R. (1947) 'On the writing of speculative fiction'. In: Latham, R. (ed.) *Science Fiction Criticism: An Anthology of Essential Writings*. New York and London: Bloomsbury Academics, 1947, pp. 17–21.

Hoad, P. (2013) '*The Impossible* is oblivious to the sea-change in mainstream global cinema'. *Al Jazeera* Weblog. [Online] Available from:. www.aljazeera.com/indepth/opinion/2013/02/201325101227523538.html Accessed 11 April 2020.

Iveković, R. (2019) *Politiques de la traduction. Exercices de partage*. Marseille: TERRA-HN éditions Alterego.

Kallis, G., Demaria, F. and D'Alisa, G. (2015) 'Introduction: degrowth'. In: Demaria, F., D'Alisa, G., and Kallis, G. (eds) *Degrowth: Vocabulary for a New Era*. London: Routledge, pp. 1–17. https://doi.org/10.4324/9780203796146

Kant, I. (1756) *History and Physiography of the Most Remarkable Cases of the Earthquake which Towards the End of the Year 1755 Shook a Great Part of the Earth*. Hong Kong: Philopschy (this edition 1994).

Kant, I. (1788) *Critique of Practical Reason*. London: Longman (this edition 1967).

Keys, D. (2000) *Catastrophe. An Investigation into the Origins of the Modern World*. New York: Ballantine Books.

Kline, W. (2005) *Building a Better Race: Gender, Sexuality, and Eugenics from the Turn of the Century to the Baby Boom*. Berkeley and Los Angeles: University of California Press.

Koppenfels, von W. (2004) '"These irritant bodies": Blinding and blindness in dystopia'. *Cambridge Quarterly* 33 (2), pp. 155–172. http://doi.org/10.1093/camqtly/33.2.155

Kothari, A., Salleh, A., Escobar, A., Demaria, F. and Acosta, A. (eds) (2019) *Pluriverse. A Post-Development Dictionary*. New Delhi: Tulika Press. https://doi.org/10.1016/j.ecolecon.2017.08.012

Lapierre, D. (2002) *Five Past Midnight in Bhopal*. New York: Warner Books.

Latour, B. (2018) *Down to Earth: Politics in the New Climatic Regime*. Cambridge, UK and Medford, MA: Polity Press.

Leonardi, E. (2017) 'For a critique of neoliberal green economy: A Foucauldian perspective on ecological crisis and biomimicry'. *Soft Power* 5 (1), pp. 169–185. http://doi.org/1017450/170110

Leonardi, E. (2019) 'The topicality of André Gorz's political ecology'. In: Barca, S., Chertkovskaya, E. and Paulsson, A. (eds) (2019) *Towards a Political Economy of Degrowth*. London: Rowman & Littlefield, pp. 55–68.

Leong, D. (2016) 'The mattering of Black Lives: Octavia Butler's hyperempathy and the promise of the new materialisms'. *Catalysts* 2 (2), pp. 1–35. https://doi.org/10.28968/cftt.v2i2.28799

Levine, P. and Bashford, A. (2010) *The Oxford Handbook of the History of Eugenics*. New York: Oxford University Press.

Lewis, S. L. and Maslin, M. A. (2015) 'Defining the Anthropocene'. *Nature* 519, pp. 171–180. https://doi.org/10.1038/nature14258

Linebaugh, P. and Rediker, M. (2000) *The Many-Headed Hydra: Sailors, Slaves, Commoners, and the Hidden History of the Revolutionary Atlantic*. Boston: Beacon Press.

Lloro-Bidart, T. and Semenko, K. (2017) 'Toward a feminist ethic of selfcare for environmental educators'. *The Journal of Environmental Education* 48 (1), pp. 18–25. https://doi.org/10.1080/00958964.2016.1249324

Lob, J., Legrand, B. and Rochette, J. (1982) *Le Transperceneige*. Tournai: Casterman.

Lohan, T., Hazen, D., Parramore, L., Geneva, T., Kelley, L., Escobedo Shepherd, J., Gwynne, K., Seltzer, S. M., Turse, N., Khalek, R., Stan, A. M., Holland, J. and Jaffe, S. (eds) (2011) *The 99%: How the Occupy Wall Street Movement Is Changing America*. San Francisco: AlterNet Books.

Lorde, A. (1988) *A Burst of Light and Other Essays*. New York: Ixia Press (this edition 2017).

Lugones, M. (2011) 'Methodological notes toward a decolonial feminism'. In: Isasi-Díaz, A. M. and Mendieta, E. (eds) *Decolonizing Epistemologies. Latina/o Theology and Philosophy*. New York: Fordham, pp. 68–86.

Luke, T. W. (2018) 'Tracing race, ethnicity, and civilization in the Anthropocene'. *Environment and Planning D: Society and Space* 38 (1), pp. 129–146. https://doi.org/10.1177/0263775818798030

Malavasi, G. (2018) *Manfredonia. Storia di una catastrofe continuata*. Milan: Jaca Book.

Marx, K. (1844) 'Economical and Political Manuscripts of 1844'. In: Elster, J. (ed.) *Karl Marx. A Reader*. Cambridge: Cambridge University Press (this edition 1986).

Masters, R. D. and Kelly, C. (eds) (1992) *The Collected Writing of Rousseau*. Vol. 3. Hanover, NH: The University Press of New England.

McCarthy, C. (2006) *The Road*. New York: Vintage Books.

McFadden, P. (2020) Contemporarity: sufficiency in a radical African feminist life. [Lecture at CES – University of Coimbra 27 January].

McKittrick, K. (2013) 'Plantation futures'. *Small Axe: A Caribbean Journal of Criticism* 17 (3), pp. 1–15. http://doi.org/10.1215/07990537-2378892

Merchant, C. (1980) *The Death of Nature: Women, Ecology, and the Scientific Revolution*. London: Wildwood House.

Merchant, C. (1989) *Ecological Revolutions: Nature, Gender, and Science in New England*. Chapel Hill, NC: University of North Carolina Press.

Mercier-Faivre, A. and Thomas, C. (eds) (2008) *L'invention de la catastrophe au XIIIe siècle. Du châtiment divin au désastre naturel*. Genève: Droz.

Merril, J. (2017) 'What do you mean: science? Fiction?' In: Latham, R. (ed.) *Science Fiction Criticism: An Anthology of Essential Writings*. New York and London: Bloomsbury Academics, pp. 22–36.

Michaeli, I. (2017) 'Self-care: an act of political warfare or a neoliberal trap?' *Development* 60, pp. 50–56. https://doi.org/10.1057/s41301-017-0131-8

Mies, M. (1986) *Patriarchy and accumulation on a world scale: women in the international division of labour*. London: Zed Books.

Mignolo, W. D. (2000) *Local Histories/Global Designs*. Princeton: Princeton University Press.

Mignolo, W. D. and Walsh, C. E. (2018) *On Decoloniality: Concepts, Analytics, Praxis*. Durham, NC: Duke University Press. https://doi.org/10.1215/9780822371779

Milanez, F. and do Santos Pinto, M. (2017) 'Nego fugido y la rebelión esclava en el antropoceno'. *Ecología Política. Cuadernos de debate internacional* 53, pp. 72–75.

Mirzoeff, N. (2018) 'It's not the Anthropocene, it's the white supremacy scene, or, the geological color line'. In: Grusin, R. (ed.) *After Extinction*. Minneapolis: University of Minnesota Press.

Mitman, G., Haraway, D. and Tsing, A. L. (2019) 'Donna Haraway and Anna L. Tsing reflect on the Plantationocene'. *Edge Effects*. https://edgeeffects.net/haraway-tsing-plantationocene/

Moore, J. W. (2015) *Capitalism in the Web of Life: Ecology and the Accumulation of Capital*. New York: Verso.

Moore, J. W. (2017) 'The Capitalocene, Part I: On the nature and origins of our ecological crisis'. *The Journal of Peasant Studies* 44 (3), pp. 594–630. https://doi.org/10.1080/03066150.2016.1235036

Moore, J. W. (2018) 'The Capitalocene, Part II: accumulation by appropriation and the centrality of unpaid work/energy'. *The Journal of Peasant Studies* 45 (2), pp. 237–279. https://doi.org/10.1080/03066150.2016.1272587

Moore, T. (1516) *Utopia*. London and New York: Penguin Books (this edition 2003).

Nancy, J. L. (1996) *Being Singular Plural*. Stanford, CA: Stanford University Press (this edition 2000).

Peet, R. and Watts, M. J. (2004) *Liberation Ecologies: Environment, Development and Social Movements*. London and New York: Routledge. https://doi.org/10.4324/9780203235096

Peet, R., Robbins, P. and Watts, M. J. (eds) (2010) *Global Political Ecology*. London and New York: Routledge.

Plumwood, V. (1993) *Feminism and the mastery of nature*. London, Routledge.

Plumwood, V. (2002) *Environmental culture: the ecological crisis of reason*. London, Routledge.

Pogliano, C. (1984) 'Scienza e stirpe. Eugenetica in Italia'. *Passato e presente* 5, 61–97.

Pogliano, C. (2005) *L'ossessione della razza: antropologia e genetica nel XX secolo*. Pisa: Edizioni della Normale.

Povinelli, E. (2016) *Geontologies. A Requiem to Late Liberalism*. Durham and London: Duke University Press. https://doi.org/10.1215/9780822373810

Pulido, L. (2018) 'Racism and the Anthropocene'. In: Mitman, G., Armiero, M. and Emmett, R. S. (eds) *Future remains*. Chicago: University of Chicago Press, pp. 116–128.

Rousseau, J. J. (1755) *Discours sur l'origine et les fondements de l'inégalité parmi les homes*. Paris: Gallimard (this edition 1966).

Rozario, K. (2007) *The Culture of Calamity. Disaster and the Making of Modern America*. Chicago: University of Chicago Press.

Sagan, D. (2017) 'Coda. Beautiful monsters: terra in the Cyanocene'. In: Tsing, A. L., Swanson, H., Gan, E. and Bubandt, N. (eds) *Arts of Living on a Damaged Planet: Ghosts and Monsters of the Anthropocene*. Minneapolis: University of Minnesota Press, pp. M169–M174.

Saldanha, A. (2017) 'Globalisation as a crisis of mobility: a critique of spherology'. In: Baldwin, A. and Bettini, G. (eds) *Life Adrift: Climate Change, Migration, Critique*. London and New York: Rowman & Littlefield International, pp. 151–173.

Salerno, D. (2016) 'Risky subjects in time of terror. A semiotic perspective on the security discourse in Europe'. *Versus* 123 (2), pp. 363–384. http://doi.org/10.14649/85508

Sassen, S. (1991) *The Global City: New York, London, Tokyo*. Princeton, Princeton University Press (this edition 2001).

Sassen, S. (1994) *Cities in a World Economy*. Thousand Oaks, CA: Pine Forge Press (this edition 2011).

Sharpe, C. (2009) *Monstrous Intimacies: Making the Post-Slavery Subject*. Durham: Duke University Press. https://doi.org/10.1215/9780822391524

Simmel, G. (1903; 1908) 'Space and the spatial ordering of society'. In: Simmel, G. *Sociology. Inquiries into the Construction of Social Forms*. Volume 1. Leiden and Boston: Brill, pp. 543–621 (this edition 2009).

Snodgrass, M. E. (2008) *The Underground Railroad: An Encyclopedia of People, places, and Operations*. Armonk, NY: M.E. Sharpe.

Sousa Santos, B. (2014) *Epistemologies of the South. Justice against Epistemicide*. London: Paradigm. https://doi.org/10.4324/9781315634876

Spiro, J. P. (2008) *Defending the Master Race: Conservation, Eugenics, and the Legacy of Madison Grant*. Burlington: University of Vermont Press.

Stannard, D. E. (1992) *American Holocaust: Columbus and the Conquest of the New World*. Oxford: Oxford University Press.

Stepan, N. (1982) *The Idea of Race in Science: Great Britain 1800–1960*. London: MacMillan.

Stephens, B. and Sprinkle, A. (2011) 'Ecosex Manifesto 1.0'. Available at: http://sexecology.org/research-writing/ecosex-manifesto/

Stephens, B. and Sprinkle, A. (2016) 'Ecosexuality'. In: Van Der Tuin, I. (ed.) *Gender: Nature*. London: Macmillan, pp. 313–330.

Stoler, A. L. (2016) *Duress. Imperial Durabilities in Our Times*. Durham: Duke University Press. https://doi.org/10.1215/9780822373612

Swyngedouw, E. and Ernstson, H. (2018) 'Interrupting the anthropo-obscene: immuno-biopolitics and depoliticizing ontologies in the Anthropocene'. *Theory, Culture & Society* 35 (6), pp. 3–30. https://doi.org/10.1177/0263276418757314

Tagliapietra, A. (ed.) (2004) *Sulla catastrofe. L'illuminismo e la filosofia del disastro*. Milan: Bruno Mondadori.

Tola, M. (2018) 'Between Pachamama and Mother Earth: gender, political ontology and the rights of nature in contemporary Bolivia'. *Feminist Review* 118 (1), pp. 25–40. https://doi.org/10.1057/s41305-018-0100-4

Tola, M. (2019) 'Planetary lovers: on Annie Sprinkle and Beth Stephens's *Water Makes Us Wet*'. In: Ferdinand, S., Villaescusa-Illán, I. and Peeren, E. (eds) *Other Globes*. London: Palgrave, pp. 231–248. http://doi.org/10.1007/978-3-030-14980-2

Tsing, A. L. (2015) *The Mushroom at the End of the World: On the Possibility of Life in Capitalist Ruins*. Princeton: Princeton University Press.

Tuana, N. (2008) 'Viscous porosity. Witnessing Katrina'. In: Alaimo, S. and Hekman, S. (eds) *Material Feminisms*. Durham: Duke University Press, pp. 188–213.

Velescu, E. (2017) *La catastrophe naturelle en littérature et peinture au XVIIIe siècle dans l'espace franco-allemande*. Paris: EPHE Sorbonne.

Vergès, F. (2017) 'Racial Capitalocene'. In: Johnson, G.T. and Lubin, A. (eds) *Futures of Black Radicalism*. London and New York: Verso, pp. 72–82.

Verne, J. (1864) *Journey to the Centre of the Earth*. Ware: Wordsworth Editions Ltd. (this edition 1996).

Voltaire. (1756) 'Poem on the Lisbon's Disaster'. In: Voltaire. *Mélanges*. Paris: Gallimard (this edition 1961).

Voltaire. *Candide: or, All for the Best*. Peterborough, ON: Broadview, 2009 [1759].

Walsh, C. E. (2015) 'Life, nature and gender otherwise: feminist reflections and provocations from the Andes'. In: Harcourt, W. and Nelson, I. L. (eds) *Practising Feminist Political Ecologies: Moving beyond the 'Green Economy'*. London: Zed Books, pp. 101–130.

Wekker, G. (2016) *White Innocence. Paradoxes of Colonialism and Race*. Durham: Duke University Press. https://doi.org/10.1215/9780822374565

Wright, W. J. (2018) 'As above, so below: Anti-Black violence as environmental racism'. *Antipode*. https://doi.org/10.1111/anti.12425.

Wyndham, D. (2003) *Eugenics in Australia: Striving for National Fitness*. London: The Galton Institute.

Wynter, S. (1971) 'Novel and history, plot and plantation'. *Savacou* 5, pp. 95–102.

Yusoff, K. (2018) *A Billion Black Anthropocenes or None*. [ebook reader] Minneapolis: University of Minnesota Press.

Zazzara, G. (2009) *Il petrolchimico*. Padova: Il Poligrafo.

Films

Amiel, Jon. *The Core*. USA, 2003.

Bay, Michael. *Armageddon*. USA, 1998.

Bayona, Juan Antonio. *The Impossible*. USA, 2012.
Blomkamp, Neill. *Elysium*. USA, 2013.
Cameron, James. *Terminator*. USA, 1984.
Cameron, James. *Terminator 2: The Judgement Day*. USA, 1991.
Cameron, James. *Avatar*. USA, 2009.
Cuarón, Alfonso. *The Children of Men*. USA, UK and Japan, 2006.
Darabont, Frank. *The Mist*. USA, 2007.
Emmerich, Roland. *The Day After Tomorrow*. USA, 2004.
Forster, Mark. *World War Z*. USA and Malta, 2013.
Garland, Alex. *Annihilation*. USA, 2018.
Gwo, Frant. *The Wandering Earth*. China, 2019.
Hillcoat, John. *The Road*. USA, 2009.
Joon Ho, Boong. *Snowpiercer*. USA, 2013.
Lawrence, Francis. *I am Legend*. USA, 2007.
Leder, Mimi. *Deep Impact*. USA, 1998.
Lee, Spike. *When the Levees Broke. Requiem in Four Act*. USA, 2006.
Malick, Terrence. *The Thin Red Line*. USA, 1998.
McCarthy, Colm. *The Girl with All the Gifts*. UK, 2016.
McG. *Terminator Salvation*. USA, 2009.
Meirelles, Fernando (2008), *Blindness*. USA and Brazil, 2008.
Miller, Tim. *Terminator: Dark Fate*. USA, 2019.
Mostow, Jonathan. *Terminator 3: Rise of the Machines*. USA, 2003.
Nolan, Christopher. *Interstellar*. USA, 2014.
Pal, George, *The War of the Worlds*. USA, 1953.
Parker, Bradley. *Chernobyl Diaries*. USA, 2012.
Peli, Oren. *Paranormal Activity*. USA, 2007.
Spielberg, Steven. *War of the Worlds*. USA, 2005.
Sputore, Grant. *I Am Mother*. Australia and USA, 2019.
Villeneuve, Denis. *Arrival*. USA, 2016.
Winterbottom, Michael. *Code 46*. UK, 2003.

TV series

Aguilera, Pedro. *3%*. Brazil, 2016– (three seasons).
Copus, Nick. *Ice 2020*. USA, 2011.
Darabont, Frank. *The Walking Dead*. USA, 2010– (ten seasons).
Mitchell, Dominic and Jonny Campbell. *In the Flesh*. UK, 2013–15 (two seasons).

Exhibitions

Adam, Rhiannon. *Big Fence/Pitcairn Island*. Photographers Gallery, London, 2019.
OMA, Alejandro Aravena, Toyo Ito and Shigeru Ban. 'Creation from Catastrophe exhibition'. The Architecture Gallery, London, 2015.

Conclusions

1 COVID-19: chronicles of the 'we' in the time of the pandemic (March 2020)

1.1 The virus spreads where the Anthropocene wastes

> Increasing evidences demonstrate the link between the 2019-nCoV and other similar known coronaviruses (CoV) circulating in bats, and more specifically those of the Rhinolophus bat sub-species. These sub-species are abundant and widely present in Southern China, and across Asia, the Middle East, Africa and Europe. Recent studies indicate that [...] humans exposure to bat-CoVs might be common. However, the route of transmission to humans at the start of this event remains unclear. Bats are rare in markets in China but hunted and sold directly to restaurants for food. The current most likely hypothesis is that an intermediary host animal has played a role in the transmission. Identifying the animal source of the 2019-nCoV would help to ensure that there will be no further future similar outbreaks with the same virus and will also help understanding the initial spread of the disease in the Wuhan area. It would also increase our understanding of the virus and help us understand how these viruses jump from animals to humans. Thus, providing critical knowledge on how to protect us from future similar events. In this regard, strengthening food control and market hygiene activities in live food market will be essential to protect people from similar and other zoonotic diseases.[1]
>
> (WHO, Situation Report 22, 11 February 2020)

Inspired by the 2002 SARS pandemic, *Contagion*, by Steven Soderbergh (USA, 2011), follows the story of a deadly virus outbreak. The final scenes reveal how the contagion started because of deforestation in China. Wild animals (bats) are forced out of their habitat and migrate to a pig farm in Hong Kong, where a piglet eats a piece of fruit dropped by a bat. In the next scene, a Chinese chef is getting ready to cook the infected piglet and leaves the kitchen to greet his customers without washing his hands. The ending seems to imply that a lack of hygiene (in 'non-civilised' places) is among the culprits of zoonotic spillover, where viruses jump from animals to humans, beside intensive meat production, the sale of live animals at wet markets, and disruptions to ecosystems caused by rapid urbanisation.[2]

1.2 The last men on Earth

> I am queuing outside the supermarket. The decree the Italian government approved last night (11 March 2020) requires us to keep a distance of at least 1 metre from one another. We can go in two at a time. There is no one around except the police. Children are not allowed to queue with their family, nor can they enter the market. We are only allowed to go to the nearest one. Once in, staff show me the yellow taped crosses on the floor where I need to stop to keep a safe distance from the customer ahead of me and the one behind me. We can't go anywhere else. Only a few shops are open at present (supermarkets and pharmacies, and a couple of stores that sell essentials). All the other stores were shut down by decree. We must carry a self-certification every time we leave the house, or else we'll be fined.[3]
>
> (12 March 2020)

In *Contagion* (2011), the government oversees food distribution. People wait in an orderly queue as a cordon of police stands guard, but riots break out as soon as the food runs out. In the meanwhile, those who managed to get in need to leave their groceries, masks and gloves behind when they bump into the infected. In *Blindness* (2009), the blind are confined to an

Figure C.1 Still from *Blindness* (2008), directed by Fernando de Meirelles showing the protagonist (Julianne Moore) in search for food at a supermarket in São Paulo.

Figure C.2 Azzurra Menzietti, *Empty shelves in a supermarket in Pisa, Italy*, 24 February 2020.

abandoned mental hospital patrolled by the military. Soldiers leave food rations at the hospital gates. Only two people are allowed to go out and collect them. Transgressors will be shot to death. Once the protagonist flees the hospital with her fellow blind inmates, she enters a supermarket in search of food, but there is nothing left to take.

1.3 Under siege

In the days of self-isolation and social distancing, a Swiss hotel's slogan 'hotel service without the other people' has taken on a whole new meaning. Because the hotel chain Le Bijou was already set up so guests could avoid human contact, its co-founder and chief executive Alexander Hübner told Insider that it didn't take much to launch 'COVID-19 Service'. The service can only be described as a luxury COVID-19-themed package with options for doctor visits and food delivery. Le Bijou has 42 units – which Hübner describes as 'luxury Airbnbs' – at properties in cities across Switzerland. Units have full-service kitchens and amenities like saunas, Jacuzzis, fireplaces, and gyms, according to Hübner. [...] With tourism and events having dropped sharply, so have the prices at Le Bijou: What would usually cost $1,000 a night is currently being offered for $500. Guests can have a health professional check in on them twice a day for an additional $800 a day or take an in-room coronavirus test for another $500. A 24/7 nurse service is also available for an extra $5,000 a day, according to Hübner. A two-week stay can therefore cost anywhere between $6,000 and $77,500, should guests opt for the 24/7 nurse service and a coronavirus test. [...] USA Today reports that hotels in Thailand and Australia are offering similar coronavirus-themed packages.[4]

(*Insider*, 25 March 2020)

We are barricaded in our homes, segregated from everyone else. We try to live a normal life, we host virtual dinners and online meetings. We pass time on the phone, we read and we listen to the news. Basically, it feels like an Orwellian tale. We live a disciplined life – government institutions, the news and social media keep us updated. Luckily, some of us can work from home. We have plenty of fresh and canned food, all the trappings of a comfortable lifestyle. Just like in Cold War times, or alien films, we stay in our shelters, our very own ivory towers, while contagion is striking down out there, despite all precautions. Out there, that's where the monsters are: we're in the red zone. And yet we lack all the basics, like hand sanitiser, face masks and reliable information. Meanwhile, being forced to live under the same roof is taking its toll on us, it makes us aggressive and anxious, as if fear was bringing out the worst in us.[5]

(19 March 2020)

I've been home alone with my dog since the quarantine started. My daughter studies in the Netherlands. The outbreak has just started there. I'm scared, I'm lonely, and if it weren't for the dog I would not know how to cope.[6]

(8 March 2020)

We all live under the same roof, my family and his. There are eight of us in three rooms. It can be fun at times. We thought we'd rather go through this together, no matter what. At least we can use family money, we have no income at the moment. But it feels like an overcrowded prison.[7]

(21 March 2020)

My brother suffers from mental disability. Captivity is killing him – and me.

(25 March 2020)

Last night, a homeless man was fined for sleeping in the street, the time is not far when they will be arrested for being outside.[8]

(11 March 2020)

Figure C.3 Still from *28 Days Later* (2003), directed by Danny Boyle, showing the protagonist Jim (Cillian Murphy) crossing Westminster Bridge to an empty London.

> Until a few weeks ago, she would drop her 22-month-old son off at a day care facility on her way to work at a nursing home in Cedar Grove, N.J. But the center has closed temporarily amid the pandemic, leaving [me] with no choice but to skip work when she can't find a friend or relative to care for him. 'This week I called out twice', Ms. Dagrin said Wednesday. 'They called me and said: "We won't accept no more call-outs. If you call out again you're out of a job"'. She said she didn't know what she was going to do for the rest of the week.[9]
>
> (*New York Times*, 28 March 2020)

In *The Walking Dead* (2010–), self-managed walled communities grow their own food or steal from others, seizing guns and all sorts of weapons from the 'enemy', barricading themselves into a fortress from where monsters and other humans are excluded. In *The Mist* (2009), survivors are trapped in a supermarket with plenty of supplies to keep them going. But forced coexistence in a confined space creates monsters. In *I am Legend* (2007), the protagonist lives with his dog in his fortress-like house, apparently the last man on Earth.

1.4 I cannot touch you

Quarantine means that people cannot touch each other. You can't hug your children if they do not live with you, and even if they do, better not tempt fate. You can't hold your partner's hand without wearing

> gloves and a face mask. And what it means is that you can't see the faces that define your daily life, let alone meet strangers. Sex with a stranger is out of the question. It can take a toll on your friendships, old and new ones alike. It feels like they're turning you into a monster, or vice-versa. I wonder if this stranger terror will be the norm when the 'Day After' comes.[10]
>
> (20 March 2020)

A question is left hanging in the air in *In the Flesh* (2013–2015) when the time comes to reintegrate 'cured' zombies back into society, 'what if they go back to a feral state and kill us all?'. In *The Walking Dead* (2010–), all survivors are infected with the virus. They just turn into zombies when they die.

1.5 *Inside the 'red zone'*

> We are not allowed to travel. Our mobility is limited by law and by the police. COVID-19 is turning Italy, and soon the whole of Europe, into a constellation of feudal fiefs from where people cannot escape. The lesson from Brexit is that the desire for isolation belongs to the colonial fantasy of the last safe and civilised place on earth. Meanwhile, mobility is tantamount to a critical threat, as they don't have the means to verify whether we're in good health. Matteo Salvini [the leader of the far-right party the League] said that Europe should close its borders and protect itself from the *external* threat, as the United Kingdom did. Basically he is saying that he will do his best to let us all die from spreading the virus among us. We are the plague. Sooner or later someone will bomb us.[11]
>
> (20 March 2020)

> A Mexican border town felt like a scene straight from a film last Wednesday, when a group of Mexicans protesters attempted to prevent the entry of U.S. travellers from Arizona, for fear that the coronavirus would spread in their country. 'Stay Home' and 'We Demand Border Shutdown' read the signs carried by people wearing face masks who had temporarily blocked the toll-gate at Nogales, Arizona's main point of entry into Mexico, according to an article in *El Diario de Sonora*. [...] The protests had immediate effects: a federal judge in Sonora issued a decree on the same day, restricting all non-essential travel from the U.S. to Nogales and calling for stricter health measures at the entry point into Mexico. The measures came a week after the U.S. government's ban on all non-essential travel across U.S. borders with Mexico and Canada to stop the spread of the virus. Soon after President Trump's announcement, chaos broke out in San Ysidro (California), the world's largest border crossing [...] Building a wall to separate the U.S. and Mexico has been high on Trump's agenda since

the presidential campaign. That possibility seems remote now, due to the potential economic fallout from the health emergency.[12]

(*Univision noticias*, 23 March 2020)

For the last five years, Greece's public health system has been unimaginably stretched by an ongoing influx of refugees from Syria and the war-torn Middle East. About 150,600 displaced persons now call Greece home, living in dystopian camps on the mainland and some of the islands. The islands are already on the brink of collapse. Lesvos, for instance, hosts a refugee camp named Moria which was originally designed to temporarily house 3,000 people. Yet tens of thousands have been crammed into it, for years-long periods. The locals are furious at the Greek government's agreement with the EU to prevent refugees from leaving the islands after they arrive via boat from Turkey, on their long journey north. Last month, locals took over much of their island, blocking refugee boats from docking on the island, and attacking NGO workers and journalists. Riot police were sent by ferry from Athens to bring the chaos under control. The Turkish government has deliberately inflamed the situation. In an effort to bully the EU into supporting Turkey in its clashes with Syria and Russia, the Turks sent tens of thousands more refugees towards the border with Greece. Greek police clashed repeatedly with refugees and Turkish border guards. Vigilante groups of local residents formed militias to hunt down any refugees who got through.[13]

(*Business Insider*, 23 March 2020)

In *28 Weeks Later* (2007), the non-infected are forced to vacate the 'red zone', leaving their loved ones behind. In *World War Z* (2013), the wall surrounding Jerusalem separates the worthy and the unworthy – the 'humans' and the 'monsters'. In *The Walking Dead* (2010), the only way to stay safe is to gather in small communities ready to wage war against one other if survival is at stake. In *The Day After Tomorrow* (2004) and *Ice 2020* (2011), masses of people head south to flee the consequences of a new ice age in the northern hemisphere. The Mexican government immediately orders the closure of the border with the U.S. In *Code 46*, undocumented refugees and migrants who do not have the required papers to enter the safer walled cities are left to die outside.

1.6 As heard on TV. And other tales from social media[14]

They say it was the Chinese that brought us the virus. Crooks. They think the rules don't apply to them. Scumbags. The president of the Veneto region, [Luca] Zaia, just said on the telly that he saw the Chinese eating live mice.[15] They're just like animals themselves, and who pays the price? Us. The civilised.

(27 February 2020)[16]

> Abroad they think all Italians are virus spreaders. Same old story, we're the mafiosos, a bunch of cheats, all thieves, filthy and unable to govern ourselves. We are either too lazy or drama queens, even in declaring an emergency.[17]
>
> (28 February 2020)

> Someone is still saying that the boat people landing in Europe are the ones to blame.[18]
>
> (28 February 2020)

In *The Horde* (2009), it is the internal and postcolonial abject from the Paris banlieues who are turned into zombies and assault the city. In *World War Z* (2013), the wall around Jerusalem manifestly separates the living from the zombie-ridden Out there, Palestine.

1.7 *The unwitting spreaders are young and run fast*

> On Friday night (March 6), there were hundreds of people partying in the club district of Porta Venezia, Milan. Just saying. Some bars, even though only table service is allowed, were happily selling take away beers and drinks. Others instead followed the rules: numbered entrance tickets, up to the number of available seats (obviously disregarding the 1-metre distance rule), and door staff to keep people out. The bottom line is, there were plenty of crowds, with dozens of small groups hanging out in the streets around Via Lecco. All you needed to do was head over to Corso Como or Largo La Foppa and you knew that the city has no intention of giving up its weekend cocktails, even if the emergency is getting worse. At 1 a.m. in the morning, thousands of youngsters, the same ones that were supposed not go to class the next day due to the government's decision to close all schools, headed to the clubs between Garibaldi and Moscova. Hugs, kisses, handshakes – forget social distancing, despite the guidelines issued by the government and health officials. Same situation at the Navigli, where 'the clubs were bursting with people'.[19]
>
> (*Il fatto quotidiano*, 7 March 2020)

In *28 Weeks Later* (2007), the asymptomatic carrier does not want to know that they can be a monster, too. In *World War Z* (2013), inside the walls of Jerusalem survivors cheer, sing, hug, and feel safe. Their noise attracts the rabid monsters, who climb over a pile of bodies to jump over the wall into the citadel and exterminate them. In *The Impossible* (2011), the holidaying tourists find out what it means to live a precarious life only when the security fence around their resort in Thailand is pulverised by the tsunami. In *Monsters*, the protagonists find out that the aliens have disrupted all the borders and walls that would allow them to go back to their safe and privileged world.

1.8 *Neoliberalism, social Darwinism and the pandemic*

In Tennessee, people with spinal muscular atrophy will be 'excluded' from intensive care units. In Minnesota liver cirrhosis, lung disease and a weak heart will deny COVID-19 patients the right to ventilatory support. In Michigan, priority is given to those who work in essential services. In Washington state, which was the first to get hit, as well as in the states of New York, Alabama, Tennessee, Utah, Minnesota, Colorado and Oregon, doctors are called to assess levels of physical and cognitive ability before deciding whether a life should be saved. [...] Of the approximately 36 states that have disclosed their criteria, a dozen list considerations regarding cognitive faculties, and others mention specific conditions that may lead to discrimination against people with disabilities. That's most evident in Alabama. In a document called *Scarce Resource Management*, it is stated that 'mentally disabled people are improbable candidates for breathing support'. Advocacy groups for people with disabilities have also raised concerns about Washington state's guidelines mentioning 'cognitive abilities' as well as about Maryland's and Pennsylvania's reference to 'severe neurological disorder'.[20]

(*L'Avvenire*, 25 March 2020)

The Spanish military has found older residents of some care homes 'completely abandoned' and even 'dead in their beds', Defence Minister Margarita Robles said in a television interview on Monday. They were found as soldiers disinfected and provided emergency health care services this week to residential homes across the country. Robles did not give an exact figure for the number of dead bodies found by Spanish soldiers.[21]

(*NPR* [National Public Radio], 24 March 2020)

They sent us to the frontline without weapons or ammunition. We are still using the FFP2 masks, the ones without filters. At the hospital, we have but a few hazmat suits. A few days ago, we ran out of hand sanitiser, and healthcare workers were working without masks: we found ourselves in this surreal situation where the patients' relatives were coming in wearing masks and we did not have any. After years of healthcare cuts from the government, we are left not only without means but also without personnel able to cope with an emergency of this magnitude. Now we are all paying the price, especially patients, who are paying with their lives. [...] Under orders of the powers that be, we are not supposed to intubate patients over 80. In short, we choose who lives and who doesn't. They are parents, husbands, and fathers. And that's the hardest part of this job. I am a nurse, and to see someone who is in pain and not be able to help them makes me feel sick. And if the numbers of the infected continue to rise, the cut off age for intubation will drop.[22]

(*Open*, 13 March 2020)

In *The Day After Tomorrow* (2004), the most vulnerable, like the scientists trapped on a remote moor in Scotland, know that their time has come. In *Ice 2020* (2011), a father with cancer chooses to sacrifice himself to let his family out of the immigration detention centre where they are held. In *Blindness* (2009), the protagonist, who is imumne to the virus, leads us inside the abandoned mental hospital where the blind have been gathered in increasing numbers and left suffering with no medical care. In *The Road* (2009), the most vulnerable are kidnapped by cannibals and kept as food supply in the basement of an empty house. In *I Am Mother* (2019), the unworthy are left outside the fortified base by 'Mother' and exterminated by the robotic soldiers she leads.

1.9 The vulnerable within us

They will first save the young and the healthier, I heard. I have cancer. I almost have a heart attack when someone sneezes. I'm running out of chemo drugs, and I can't find them in pharmacies. On the other hand, maybe I should take a break from treatment, to strengthen my immune system.[23]

(16 March 2020)

I am HIV-positive and, well, I don't go unnoticed. With such a shortage of beds and medication, if I end up in hospital with COVID 19, how many people will be ahead of me?[24]

(14 March 2020)

A 17-year-old whose death was initially linked to the novel coronavirus despite not having any previously reported health conditions was denied treatment at a California medical facility over his lack of insurance, according to the mayor. R Rex Parris, the mayor of Lancaster, California, confirmed the teen's death in a video posted to YouTube on Wednesday, in which he warned residents to take the global pandemic seriously and practice self-isolation and social distancing measures. 'The Friday before he died, he was healthy', the mayor said about the teenager. 'By Wednesday, he was dead'.[25]

(*The Independent*, 27 March 2020)

I am in the South, where the haemorrhage of youth has been going on for centuries now. Cities and villages are usually deserted for most of the year, except July and August. But today, in the times of the pandemic, every other shop and bar is closed, and the elders are left on their own. Their lives and social lives as they are, are made impossible: there is no glass of wine, no game of cards, company or mutual support. They are left in silence, with their *vulnerability* and death.[26]

(13 March 2010)

What about the undocumented migrant working in the fields in southern Italy, living in a slum, with limited access to hospitals and health care, spatially segregated and subjected to harassment by landowners and companies? And what about the undocumented prostitute who works in the hidden alleys of the North and the South and is exposed to all kinds of viruses, and who has not eaten since March 1st? And the homeless woman who doesn't know who to beg from because there is no one around? And the street vendor? How will she survive? What's their place in the hierarchy of priorities?[27]

(12 March 2020)

These figures are almost absent in the films I have analysed.

1.10 *A cry from the margins: when the expendable revolt*

Prisoners at jails in Naples Poggioreale in the south, Modena in the north, Frosinone in central Italy and at Alexandria in the northwest had all revolted over measures including a ban on family visits, unions said. Similar scenes played out in Padova in the north and Bari, Foggia and Palermo in the south, media reports said. [...] The revolt began after relatives of the inmates held a protest outside the gates over the measures preventing them from seeing their loved ones, the 'Stampa' newspaper said. At Frosinone, south of Rome, police were called in to restore order after about 100 prisoners barricaded themselves into a section of the prison [...].[28]

(Channel News Asia, 9 March 2020)

In *Ice 2020* (2011), as the new ice age starts to hit the northern hemisphere and Great Britain is covered in ice, detained illegal migrants are left behind. In *Blindness* (2008), when the virus begins to spread among the military and the government, the blind are left alone in the carceral archipelagos where they have been confined.

1.11 *We cannot grieve you*

Inside the Bergamo cemetery, dozens of coffins fill the Ognissanti church, now an emergency mortuary storing the corpses of the community's dead after the region's two hospitals could hold no more. With funerals banned under Italy's lockdown decree, the city crematorium is set to begin operating on a new 24-hour schedule this weekend to keep up. Officials had to close the cemetery to stop the elderly from coming by bus to say last respects to fellow friends, neighbours and relatives dying at an alarming rate in the city that is now the aching heart of the battle against coronavirus outbreak that has claimed over 1,200 lives in Italy, mostly in Lombardy.[29]

(*The Telegraph*, 14 March 2020)

In *28 Days Later* (2003), Jim wakes up from a coma and leaves the hospital – all around him, piles of corpses in body bags. In *The Walking Dead* (2010–), one of the main characters becomes insane at the thought that the 'walkers' are no longer his loved ones but dangerous killers and virus spreaders. In *Monsters* (2010), nobody will grieve the loss of the little Mexican girl that Andrew and Samantha found in the forest– there is no space for mourning the wretched of the earth, much less in war times.

1.12 Not everyone can stay home

Not everyone can stay safe. My husband works for the town's maintenance department. He and his crew are out in the streets every day. Sometimes they are called in for a night shift to close the parks, because the government's lockdown measures now include all public green spaces and people are forbidden to gather there.[30]

(17 March 2020)

1.13 Chaos outside the ivory tower

We are stuck abroad. All flights to Italy have been suspended. We can't be with our loved ones. We tried booking different flights, but they were all cancelled. We have a little boy. Not being able to go back home and having to survive here, it's a catastrophe. We can't even go to the airport and ask, it's patrolled by the police and they won't let you in if you don't have a valid ticket. We thought about renting a car, but we may be arrested if we try to cross the border. Being here is like being on a different planet, so different from what our family is going through back home.[31]

(11 March 2020)

If you are given a chance at living on *Elysium* (2013), you need to forget the ones you left behind in a dying Earth and treat them as monsters endangering the well-being of the best among humankind. In *3%* (2016), once you have been through the Process and gained access to the 'Maralto', you are expected to look down on the family you left behind in the 'Continent' as if they were inferior, of a lower rank.

1.14 *Will techno-science save us?*

> Oxygen therapy is the major treatment intervention for patients with severe COVID-19. All countries should work to optimize the availability of pulse oximeters and medical oxygen systems. Mortality in those with critical illness has been reported as over 50%, thus implementation of proven critical care interventions such as lung protective ventilation should be optimized.[32]
>
> (World Health Organisation [WHO], 1 March 2020)

> From the evidence so far, the COVID-19 virus can be transmitted in ALL AREAS, including areas with hot and humid weather. Regardless of climate, adopt protective measures if you live in, or travel to an area reporting COVID-19. The best way to protect yourself against COVID-19 is by frequently cleaning your hands. By doing this you eliminate viruses that may be on your hands and avoid infection that could occur by then touching your eyes, mouth, and nose. [...] The virus is so new and different that it needs its own vaccine. Researchers are trying to develop a vaccine against 2019-nCoV, and WHO is supporting their efforts.[33]
>
> (WHO, 9 March 2020)

In *Contagion* (2011), the virus, as was the case with SARS and COVID-19, is described as being caused by overproduction and the lack of hygiene in Chinese restaurants: only the self-sacrifice of the scientist who decides to test the vaccine on herself will make it immediately available to the rest of the Earth's population. In *I am Legend* as well as in *WWZ*, the virus is the result of lab experiments and the heroes will sacrifice themselves to find a cure.

1.15 *Echoes of hope from the 'red zone'*

> We're all here, far away from our dear ones, looking for numbers, sharing information and, sometime, oversharing information of a crisis which represents a real disaster for our families, friends, country and for us. We don't know when we'll see our loved ones again. We don't know when it will be possible to hug them again. This is something that resembles a war, and it is devastating. Not everything is shit, though. Nonetheless, there is something about this planetary crisis that is generating beauty, through simple, genuine acts of solidarity that show how humanity can really be amazing at times. There is a video of people in the South of Italy who started playing and singing from different balconies, while locked up in their homes.[34]
>
> (12 March 2020)

In *3%* (2016–), two of the protagonists in the 'Continent' talk on a walkie-talkie while looking at each other through their windows. Meanwhile, covert militants of the Cause in the 'Maralto' build a radio to communicate with their fellow militants in the 'Continent'.

1.16 *A human future must come*

> Even in this fatal hour, we must acknowledge the fascinating side of this story – that someday we'll be able to tell it (at least, we hope so). And we'll evoke it in a heavy tone, with that mixture of protagonism and extraordinariness that accompanies the memory of exceptional times. We will report back, albeit with exaggerated emphasis, on every detail of what we did, step by step, in those days that troubled the world. We will try to make even the ordinary sound incredible. That is the genesis of new stories.[35]
>
> (12 March 2020)

In *The Core* (2003), the slowing rotation of the Earth's core is restarted by scientists and astronauts: in the future, when the Earth's population will have been transferred to another galaxy, humans will tell the story to their children and the children of their children. In *The Day After Tomorrow* (2004) and *Ice 2020* (2011), the new ice age triggered by climate change gives birth to a new Anthropocene. Humans are safe, and the 'we' still has control over the world. In the last scene of *The Wandering Earth* (2019), the son of the astronaut who saved the world from catastrophe recounts in a hopeful tone the stages of the Earth's migration to another planet, a process that will take 2,500 years and 100 generations to complete. In *The Girl with All the Gifts* (2016), once the tree has spread its spores, every human being dies or metamorphoses into a tree. Only the hybrids – who survive the mutation – are left to tell that story.

2 The end of the word as the 'we' knows it?

In the apocalyptic scenario of COVID-19, narratives of the catastrophe have re-centred Europe and the West as the last bastion of civilisation, socio-political order and techno-scientific progress that must be preserved at all costs. The 'imagined community' of the 'we', a symbolic and material construction, has solidified around the perceived threat of danger posed by migrant 'invasions', terror attacks and environmental catastrophe, claiming to be indispensable to the survival of the human species in the 'After'. As in a dystopian memory of the future, the 'we' blamed the monster. The Chinese were the first target of *moral panic*,[36] soon followed by anyone who did not comply with the measures restricting mobility – the virus spreaders, often reported to authorities.[37] The president of the European Commission, Ursula von der Leyen, promptly announced that 26 member states had agreed to close European borders to the monster Out there.[38]

Nevertheless, inside and outside the 'red zone', some were acting as if they were untouchable, because catastrophes are meant 'for' the Other who lives in the place *for* disaster. The 'we' did not fear for its future just yet, and the moral obligation to protect the most vulnerable in society soon gave way to the normalisation of the crisis and utilitarian arguments: 'someone has to be sacrificed', like the migrants and refugees living in the confined spaces of 'reception' centres Out there, where access to healthcare and social distancing are not an option.[39] In any case, the 'we' still believed that techno-science would eventually save the 'best part of humanity'.

It has, following traditional patterns and well-established hierarchies and priorities that set boundaries and borders by semiotically defining what *humanness* is about. By the same token, the Other Out there was the least priority, although the boundary was in fact blurred (as the nurse says in entry no. 8 to the Chronicle: 'the age cut off will drop' and those who had privileged access to healthcare were soon left behind). Internal monsters were next to be sacrificed – the poor, the undocumented, the racialised, the sick, the ordinary criminals, against whom Anthropocenic necropolitics operates (see entries no. 1.9 and no. 1.10). Malthusianism combined with new forms of social Darwinism was applied to select the less vulnerable from the masses of the less fit – not only the elder and persons with disabilities, but also those precarious workers whose being is dependent on being exposed to others (e.g. sex workers, care givers, cleaners and nurses), including domestic violence perpetrators. One may have thought that the Malthusian practices conjured up in cinematic dystopian memories of the past had become a daily reality when Britain's prime minister Boris Johnson refused for weeks to implement measures to prevent the spread of the virus, claiming that 'we' needed to come to terms with the fact that many families would 'lose their loved ones ahead of time' – or when his Health secretary urged all people aged over 70 to just 'self-isolate'[40] or, again, when U.S. president Donald Trump's and Brazil's president Jair Bolsonaro's efforts seemed directed at down-playing the threat rather than protecting the vulnerable (see entry no. 1.8).

The hierarchy of priorities might change, but it seems unlikely to move away from crystallised notions of worthiness handed down through history. Authoritarian biopower has already been shaping the pandemic, relying on big data collection tools, social control and policing methods, and forms of segregation and isolation to guarantee order and security for the 'saved'.[41] And, what is worse, measures put forward in the name of public health and safety have rapidly gained legitimacy in the eyes of public opinion.

Then, if we were to imagine what the next step in such dystopian scenario might be, in a sort of reversed 'memory of the future' in which our predictions may not seem so fictional anymore, even a war suddenly does not seem a far-fetched possibility.

Yet there is nothing new in this picture. Everything has already happened outside the safe space of the 'we'. The difference lies in the fact that catastrophe in the Out there was always regarded as 'peripheral' to the 'we' and either unacknowledged or legitimised through the *monstrification* of the Other.[42]

The planet will certainly survive, and a future without humans is apparently unthinkable, but many 'Out there' will be left behind, and not by chance. New *semiotic borders* will define who is worthy and who is not, based on the 'figures of race' that since Modernity have sustained the privilege of some and legitimised the vulnerability of many. Similarly, non-human life will be sacrificed or saved depending on its usefulness to the survival of the 'we'. Because the 'we' is bound to repeat the same mistakes.

Many in Europe and the broader West are trying their best to correct the mistakes of neoliberalism, starting with the pulverisation of a collective idea of care involved in the present collapse of national health systems everywhere after decades of budget cuts. The COVID-19 emergency has shown to what degree care and self-care are interdependent. Entire communities in lockdown have created commons, singing together from their balconies, providing emotional support and free babysitting, using all sorts of digital platforms to *share* skills, knowledge, interests, films, yoga classes, recipes, even porn. In Lisbon and everywhere in the world, people have volunteered to support vendors, panhandlers, the homeless and all those whose livelihoods depend on a few coins by passers-by. Others have assisted the elderly confined to their homes by the pandemic. Many have accepted to 'stay home' not only to avoid spreading the virus but also to ease the burden on already overcrowded hospitals. Others have quarantined with

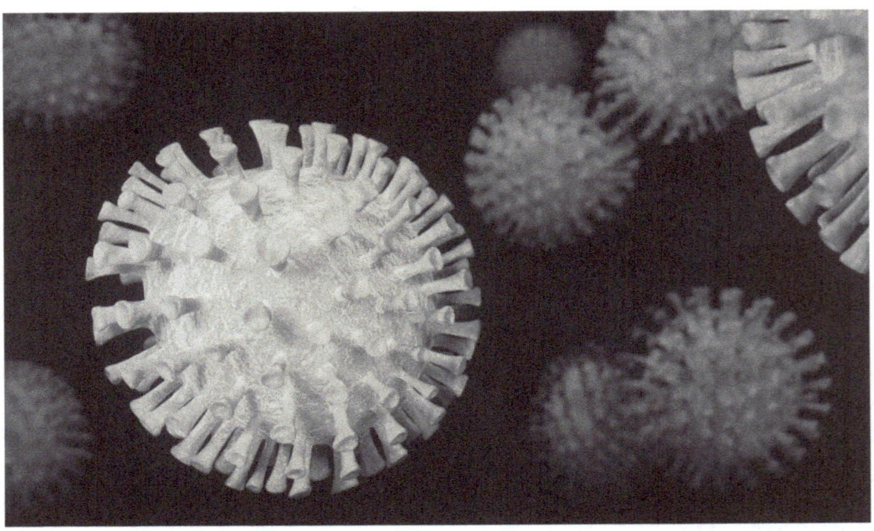

Figure C.4 Digital picture of COVID-19, https://pixabay.com/photos/coronavirus-corona-virus-pandemic-4972480/.

the more vulnerable among their friends. Hotels and innkeepers have offered to house people under mandatory quarantine orders. Some have started growing their own food, others have bought produce from small farmers. At a time when food and medication are precious commodities, many have learned to reduce waste. In New York State, the government has allowed patients to share oxygen ventilators[43] Portugal has granted all migrants and asylum seekers temporary citizenship status.[44]

With no cars on the streets and a significant decline in industrial waste and emissions (although many 'non-essential' factories were kept open), air pollution has rapidly fallen. Earth-care and fantasies of a world without value extraction were among the more immediate outcomes of the catastrophe. Workers went on strike, demanding more shutdowns and better safety measures.[45] The pandemic has made people aware that everything they touch may be a means to survival or a cause of death, revealing how human life depends on its *trans-corporeal* proximity and connection to the Earth and *intra-active* relationships with non-human Life and Nonlife: human life and subjectivity are the product of what pre-exists them, be it *bios*, *zoe* or the inorganic matter that is also a vehicle of virus transmission.

How to survive catastrophic times is probably something the 'we' should learn from these moments of shared experience, because the threat will not disappear. Maybe the 'we' should learn from these experiences of resistance without trying to reduce their local and transnational action to a question of national pride. As the 'we' plans for the future, perhaps a useful lesson might be drawn from the same alternative cosmogonies and pluriverse based on non-extractivist and non-exploitative relationships between Life and Nonlife against which the Anthropos of the Anthropocene has fought for centuries to make colonialism, slavery, capitalism and patriarchy the ruling standard. A paradigm shift in the approach towards growth and development seems to be the only way of halting the Anthropocenic race towards destruction, epitomised by the huge wet markets where the virus made the jump from animals to humans.[46] Production and distribution of goods on a smaller scale, a different allocation of energy and time and a different social organisation in line with the purposes of degrowth can no longer be postponed. Bio- and necropolitical *dispositifs* – with their borders, walls, colonies and carceral archipelagos – would mean nothing in this new paradigm, and even if they did, they would be opposed by most.

This would leave room for a deep inquiry into the role played by gender, race, sexuality and class as means of subjection, and for a chance at mitigating their impact. Similar forms of self-organisation will be found in past and present (non-hegemonic) experiences, especially where other possibilities for coexistence, multilingual and *bio*-logical communication, and *skin-to-skin partage* among beings and between beings and the Earth have emerged from situations of scarcity and emergency. My hope is that this newfound *partage* at a time of catastrophe – envisioned in films such as *Arrival* (2016), *Annihilation* (2018), *Monsters* (2010), *Under the Skin* (2013) and *The Girl with All the Gifts* (2016), and at present limited by a plethora of rules and regulations – will eventually succeed in undermining the sense of superiority of the 'we'.

But that is another film.

Notes

1 www.who.int/emergencies/diseases/novel-coronavirus-2019/situation-reports.
2 www.theguardian.com/environment/2020/mar/18/tip-of-the-iceberg-is-our-destruction-of-nature-responsible-for-covid-19-aoe.
3 'Sono in fila davanti al supermercato. Il decreto che il governo ha approvato ieri sera (11/03/2020) ci impone di stare a un metro di distanza gli uni dalle altre. Si può entrare due a due. Tutto intorno non c'è nessuno se non la polizia. I bambini non possono stare in fila con i loro parenti né entrare insieme a loro a fare la spesa. Ci è permesso di raggiungere unicamente il supermercato più vicino. Una volta dentro, lo staff del negozio mi mostra le croci gialle dove mi devo fermare per mantenere la distanza di sicurezza da chi si trova davanti a me o alle mie spalle. Non possiamo andare da nessun'altra parte. Al momento, solo alcuni esercizi son aperti (tra cui farmacie e supermercati ed altri negozi di beni essenziali) sono ancora aperti. Il decreto ha imposto a tutti gli altri di chiudere. Abbiamo bisogno di un certificato per uscire da casa, altrimenti possiamo venire denunciati con ammenda'. (Text message from a friend in Fano)
4 www.insider.com/le-bijou-hotel-switzerland-luxury-coronavirus-covid-19-package-2020-3?utm_medium=social&utm_source=facebook.com&utm_campaign=sf-bi-main.
5 'Ci siamo barricati nelle nostre case, segregati dalla vita di tutti gli altri. Cerchiamo di vivere una vita normale, organizzando cene e incontri online. Passiamo il tempo al telefono, a leggere libri e ascoltare le notizie. In pratica ci troviamo in una situazione Orwelliana, disciplinati nella nostra vita e aggiornati dalle istituzioni, i telegiornali e i social media. Alcuni di noi possono lavorare da casa, fortunatamente. C'è cibo fresco e in scatola in abbondanza, e ci godiamo tutti gli strumenti di una vita quotidiana molto confortevole. Come ai tempi della Guerra Fredda o come nei film sulle invasioni aliene, stiamo nei nostri rifugi, nelle nostre torri d'avorio private, mentre fuori la minaccia del contagio sta cacciando e colpendo anche a dispetto di ogni precauzione. Là fuori ci sono i mostri: noi siamo nella zona rossa. E tuttavia ci mancano le cose basilari. Come il disinfettante per le mani, le mascherine e informazioni certe. Nel frattempo, la convivenza forzata ci rende molto aggressivi e ansiosi, come se la paura del contagio tirasse fuori la parte peggiore di noi'. (Private Facebook message from a friend in Bologna, Italy)
6 'Sono solo con il mio cane da quando per me è iniziata la quarantena. Mia figlia studia in Olanda, dove il virus è appena arrivato. Ho paura, mi sento solo e se non fosse stato per il cane non avrei saputo come far fronte a questa situazione'. (Text message from a friend in Fano, Italy)
7 'Viviamo tutti insieme, la mia famiglia e la sua. Siamo in otto in tre stanze. È divertente a volte. Abbiamo pensato che sarebbe stato meglio affrontarlo insieme, non importa cosa. Almeno possiamo usare i soldi di famiglia, non avendo un reddito al momento. Ma è come stare in una prigione sovraffollata'. (Private Facebook message from a friend in Venice, Italy)
8 www.tgcom24.mediaset.it/cronaca/milano-denunciato-un-senzatetto-violato-il-decreto-coronavirus_16077475-202002a.shtml.
9 www.nytimes.com/2020/03/27/business/economy/coronavirus-inequality.html?action=click&module=Spotlight&pgtype=Homepage.
10 Phone call from a friend in Lisbon, Portugal. Original in English.

11 'Non possiamo viaggiare. La nostra mobilità è limitata dalla legge e dalla polizia. COVID-19 trasformerà l'Italia e poi tutta l'Europa in una manciata di feudi da cui nessuno può scappare. Brexit insegna: il desiderio di isolamento appartiene alla fantasia coloniale dell'ultimo posto sicuro e civilizzato sulla terra. Nel frattempo, la nostra mobilità è equiparata alla minaccia più grave, visto che non hanno gli strumenti per verificare le nostre condizioni. Matteo Salvini ha dichiarato che l'Europa dovrebbe chiudere i confini per proteggersi da una minaccia esterna, come ha fatto il Regno Unito. Fondamentalmente lui sta dicendo che farà del suo meglio per lasciarci morire tutti per auto-contagio. Noi siamo la peste. Prima o poi qualcuno ci bombarderà'. (Email from a friend in Milan, Italy)

12 'En una ciudad fronteriza de México se vio este miércoles una escena que solo había recreado Hollywood: un grupo de mexicanos protestó para impedir la entrada de estadounidenses desde Arizona, temiendo que continúe en su estado la peligrosa propagación del coronavirus. 'Quédate en casa' y 'Exigimos el cierre de la frontera', se leía en algunos letreros que sostenían las personas con cubrebocas que bloquearon temporalmente la garita de Nogales, la principal entrada desde Arizona hacia México, según un reporte de El Diario de Sonora. [...] La manifestación tuvo resultados inmediatos: un juez federal en Sonora concedió el mismo miércoles un amparo para que los visitantes de EEUU solo crucen hacia Nogales por viajes esenciales y pidió establecer medidas sanitarias más estrictas en ese puerto de entrada a México. Esta medida ocurre una semana después de que el gobierno de EEUU ordenara el cierre de las fronteras con México y Canadá para todos los viajes no esenciales, para detener los contagios del virus. Apenas lo anunció el presidente Donald Trump, el caos se desató en la garita de San Ysidro (California), el cruce fronterizo más grande del mundo. [...] Construir un muro que separe definitivamente a EEUU de México es uno de los proyectos que Trump ha impulsado desde que era candidato presidencial. Es una iniciativa que sigue viéndose lejana, ahora por una virtual crisis económica que se generaría por la contingencia sanitaria'. www.univision.com/noticias/america-latina/el-mundo-al-reves-mexicanos-bloquean-frontera-para-que-no-entren-a-su-pais-estadounidenses-contagiados-por-el-covid-19?fbclid=IwAR1OOs1dKvUzzAuq45Abnbsg0m0cGZteeOXGKAENPCm-60UXTTZYgEeJwQI.

13 www.businessinsider.com/coronavirus-greece-prepares-for-crisis-2020-3?IR=T.

14 Random posts on Facebook originally in English.

15 https://nypost.com/2020/02/29/politician-apologizes-for-saying-coronavirus-caused-by-chinese-people-eating-live-mice/.

16 See also a statement by Giorgia Meloni, leader of Italian far-right party Fratelli D'Italiawww.la7.it/nonelarena/video/coronavirus-meloni-i-cinesi-punto-di-riferimento-non-mi-fregano-ci-hanno-portato-il-virus-15-03-2020-313419.

17 www.wsj.com/articles/italians-are-being-treated-as-a-risk-abroad-over-coronavirus-11582913454.

18 www.theguardian.com/commentisfree/2020/feb/28/coronavirus-outbreak-migrants-blamed-italy-matteo-salvini-marine-le-pen.

19 'Venerdì sera (6 marzo), per dire, i locali in zona Porta Venezia, a Milano, erano frequentati da centinaia di persone. Alcuni bar, nonostante l'obbligo di servizio al tavolo, vendevano tranquillamente birre e bevande da asporto. Altri, invece, si sono attenuti alle regole: ingressi contingentati fino a un numero pari ai posti a sedere (ovviamente a una distanza minore di un metro) con personale di servizio alla porta per bloccare altri avventori. Il risultato, in ogni caso, è che gli 'assembramenti' c'erano, con decine di capannelli nelle strade attorno a via Lecco. Bastava poi spostarsi in Corso Como o Largo La Foppa per rendersi conto che la città non ha intenzione, nonostante l'aggravarsi dell'emergenza, di rinunciare ai cocktail del fine settimana. All'1 di notte migliaia di giovanissimi, gli stessi che il giorno successivo non avrebbero dovuto presentarsi in classe per le restrizioni

imposte dal governo sui corsi scolastici, si sono ritrovati nei locali notturni tra Garibaldi e Moscova. Abbracci, baci, strette di mano, nessuna distanza di sicurezza, nonostante le indicazioni diffuse da istituzioni e personale sanitario. Stessa situazione anche sui Navigli, dove 'i locali scoppiavano di gente'. www.ilfattoquotidiano.it/2020/03/07/coronavirus-dalla-movida-che-non-si-ferma-ai-centri-commerciali-pieni-ecco-la-milano-irresponsabile-che-ignora-le-raccomandazioni/5729026/.

20 'In Tennessee le persone affette da atrofia muscolare spinale verranno «escluse» dalla terapia intensiva. In Minnesota saranno la cirrosi epatica, le malattie polmonari e gli scompensi cardiaci a togliere ai pazienti affetti da COVID-19 il diritto a un respiratore. Il Michigan darà la precedenza ai lavoratori dei servizi essenziali. E nello Stato di Washington, il primo a essere colpito dal coronavirus, così come in quelli di New York, Alabama, Tennessee, Utah, Minnesota, Colorado e Oregon, i medici sono chiamati a valutare il livello di abilità fisica e intellettiva generale prima di intervenire, o meno, per salvare una vita. [...] Fra i circa 36 Stati che hanno reso noti i loro criteri, una decina elenca anche considerazioni di tipo intellettivo, e altri parlano di condizioni precise che possono portare alla discriminazione nei confronti dei disabili. L'Alabama è il caso più eclatante. Nel suo documento intitolato Scarce Resource Management sostiene che i 'disabili psichici sono candidati improbabili per il supporto alla respirazione'. Ma anche frasi contenute nelle linee guida di Washington, come 'capacità cognitive', o di Maryland e Pennsylvania, come 'disturbo neurologico grave', hanno suscitato l'allarme delle associazioni di difesa dei disabili'. www.avvenire.it/mondo/pagine/niente-respiratori-per-i-disabili-pi-di-10-stati-scelgono-chi-salvare?fbclid=IwAR3m90CSahjbnqu7vtK8fBAY0sCUhY9PQeKNk3gWAe1YaouqTNvmc1V7RMI.

21 www.npr.org/sections/coronavirus-live-updates/2020/03/24/820711855/spanish-military-finds-dead-bodies-and-seniors-completely-abandoned-in-care-home?t=1585206607950&t=1585389399846.

22 '[...] Siamo stati mandati al fronte senza armi e munizioni. Tutt'ora lavoriamo con le mascherine FFP2, quelle senza filtro. Abbiamo poche tute negli ospedali. Fino a pochi giorni fa nei reparti è arrivato a mancare persino l'igienizzante per le mani, i sanitari lavoravano senza mascherine: si vedeva la situazione surreale in cui i parenti arrivavano con le mascherine, mentre noi non ce le avevamo. Da anni si fanno tagli nella sanità, non ci sono mezzi, ma anche il personale per gestire un'emergenza simile. Ne stiamo pagando il conto noi, ma anche gli stessi pazienti ed è un conto molto pesante che a volte si paga anche con la vita [...] Noi abbiamo ricevuto indicazione di non intubare più i pazienti sopra gli 80 anni, in caso di emergenza. Così, in poche parole, a volte, ci si trova a scegliere chi salvare e chi lasciar vivere. Succede quando non ci sono ventilatori per tutti e posti letto per tutti. Queste persone sono padri, mariti, nonni. Ed è la parte più difficile di tutta questa situazione. Io faccio l'infermiera, vedere qualcuno che sta male e non poterlo aiutare mi sconvolge. E se i numeri delle persone contagiate continuerà a salire, l'età di intubazione si abbasserà. Torno a casa e piango ogni giorno'. www.open.online/2020/03/13/coronavirus-racconto-infermiera-combattiamo-guerra-non-siamo-preparati-difficile-scegliere-quale-vita-salvare/?fbclid=IwAR0YtWjZt8I7N2YvahtGkzKDv0BnpjpapZEnKbQS4f48P79mZehbIBlIuQM.

23 'Salveranno, come dicono, prima i giovani e tra essi quelli sani. Io ho il cancro. Ad ogni starnuto mi viene il terrore. La mia scorta di chemioterapici è in calo. Non riesco a trovarli nelle farmacie. D'altra parte, forse è meglio sospendere la terapia, per rafforzare il sistema immunitario'. (Sms from a friend in Milan)

24 'Io sono sieropositiva e come dire non passo inosservata. Se mi contagio e vado all'ospedale, quante persone dovrò aspettare perché tocchi a me essere curata?'. (Text message from a queer friend)

25 www.independent.co.uk/news/world/americas/coronavirus-teenager-death-california-health-insurance-care-emergency-room-covid-19-a9429946.html?utm_medium=Social&utm_source=Facebook&fbclid=IwAR34oAXxvYgrvR_v861nroxDzgSijCMjB4xf6njfS5FUhtc2PtAb5jBWvQA#Echobox=1585313812.
26 Original message in English. (Email from a friend in Lauria, Basilicata, Italy)
27 'E che dire del migrante senza documenti che lavora nei campi del Sud, che vive nelle baraccopoli, con difficile accesso agli ospedali e alle cure sanitarie, segregato spazialmente e sottoposto alle angherie di proprietari di terreni e di aziende? E la prostituta senza documenti che lavora in molte strade nascoste del Nord e del Sud ed è esposta a virus di ogni tipo e che dal primo di marzo non mangia più? Della senza tetto che non sa a chi chiedere l'elemosina perché in giro non c'è nessuno? E l'ambulante? Come campa quella? Qual è il loro posto nella gerarchia delle priorità?' (Private Facebook message)
28 www.channelnewsasia.com/news/world/coronavirus-covid-19-italy-prison-12516992.
29 www.telegraph.co.uk/news/2020/03/14/elderly-left-isolated-abandoned-italy-death-rate-soars/.
30 Phone call from a friend in Bologna, Italy.
31 Friends visiting in Lisbon, before the Italian government and Alitalia Airlines agreed to repatriate Italian citizens who were stuck abroad. Original in English.
32 www.who.int › 20200301-sitrep-41-covid-19.
33 www.who.int/emergencies/diseases/novel-coronavirus-2019/advice-for-public/mythbusters.
34 Original Facebook post in English. www.youtube.com/watch?v=GIEJQgNraPU&feature=youtu.be&fbclid=IwAR2hlnGtE8CEZcJSv9inRv_Mp_81BAHZSCzNkkTB9davcAaBjCxstGIS-aI.
35 'Pur nella drammaticità dell'ora fatale, dobbiamo far salvo l'aspetto affascinante della storia: e cioè che un giorno potremo raccontarla (almeno si spera). E la evocheremo in tono greve, con quel misto di protagonismo e straordinarietà che accompagnano il ricordo di tempi eccezionali, riportando ai nostri interlocutori, seppur in forma enfatizzata, ogni dettaglio di cosa facemmo, passo dopo passo, in quei giorni che sconvolsero il mondo. Proveremo a rendere incredibile anche l'ordinario. È la genesi di nuovi racconti'. (Facebook post)
36 https://en.wikipedia.org/wiki/List_of_incidents_of_xenophobia_and_racism_related_to_the_2019%E2%80%9320_coronavirus_pandemic.
37 https://time.com/5799586/italy-coronavirus-outbreak/. www.ilfattoquotidiano.it/2020/03/20/coronavirus-la-stretta-delle-regioni-in-veneto-spesa-a-200-metri-da-casa-in-sicilia-si-esce-una-volta-al-giorno-spiagge-chiuse-nelle-marche/5743144/.
38 https://news.sky.com/story/coronavirus-eus-borders-will-close-for-30-days-from-noon-tomorrow-11958658.
39 www.telesurenglish.net/news/Greece-Covid-19-Case-in-Lesbos-Spreads-Fear-Among-Migrants-20200312-0012.html. www.nytimes.com/2020/03/26/world/asia/coronavirus-refugees-camps-bangladesh.html.
40 www.thesun.co.uk/news/11158326/boris-johnson-cobra-coronavirus-2/; www.theguardian.com/world/2020/mar/15/coronavirus-uk-over-70s-to-be-asked-to-self-isolate-within-weeks-hancock-says.
41 www.theverge.com/2020/3/23/21190700/eu-mobile-carriers-customer-data-coronavirus-south-korea-taiwan-privacy.
42 While the COVID-19 pandemic was spreading all over the West and started reaching other continents, a locust invasion fed by climate change brought devastation to East Africa. www.axios.com/locust-swarms-starvation-africa-asia-65443ada-c9a1-40a8-b0b7-7c7398c1d82d.html.
43 https://edition.cnn.com/2020/03/26/health/splitting-ventilators-coronavirus/index.html.

44 https://uk.reuters.com/article/uk-health-coronavirus-portugal-idUKKBN21F0MC?utm_campaign=trueAnthem%3A+Trending+Content&utm_medium=trueAnthem&utm_source=facebook&fbclid=IwAR1mveoQuDLZ9sVWZJMe0ZbDI0BlFSxb2ZUWIbdynlOp_ghvI64e2Zi0p_k.

45 www.channelnewsasia.com/news/world/italy-s-lombardy-workers-to-strike-for-tougher-coronavirus-shutdowns-12568704. See also the reflection on grassroots struggles against Malthusianism and necropolitics by Sandro Mezzadra: www.versobooks.com/blogs/4598-politics-of-struggles-in-the-time-of-pandemic?fbclid=IwAR0jyO7Phq5ZXWyx-e-8fN9Zh49rLmUh2XT9k1xDDhbZDOclbfhB7A0HvS4.

46 www.greenqueen.com.hk/preventing-another-pandemic-the-link-between-coronavirus-and-industrial-livestock-farming/.

Films

Amiel, Jon. *The Core*. USA, 2003.
Bayona, Juan Antonio. *The Impossible*. USA, 2012.
Blomkamp, Neill. *Elysium*. USA, 2013.
Boyle, Danny. *28 Days Later*. UK and USA, 2002.
Dahan, Yannick and Benjamin Rocher. *The Horde*. France, 2009.
Edwards, Gareth. *Monsters*. UK, 2010.
Emmerich, Roland. *The Day After Tomorrow*. USA, 2004.
Darabont, Frank. *The Mist*. USA, 2007.
Forster, Mark. *World War Z*, USA and Malta, 2013.
Fresnadillo, Juan C. *28 Weeks Later*. UK and Spain, 2007.
Garland, Alex. *Annihilation*. USA, 2018.
Glazer, Jonathan. *Under the Skin*. UK, USA, Switzerland, 2013.
Gwo, Frant. *The Wandering Earth*. China, 2019.
Hillcoat, John. *The Road*. USA, 2009.
Joon Ho, Boong. *Snowpiercer*. USA, 2013.
Lawrence, Francis. *I am Legend*. USA, 2007.
McCarthy, Colm. *The Girl with All the Gifts*. UK, 2016.
Meirelles, Fernando (2008), *Blindness*. USA and Brazil, 2008.
Soderbergh, Steven. *Contagion*. USA, 2011.
Spielberg, Steven. *War of the Worlds*. USA, 2005.
Sputore, Grant. *I Am Mother*. Australia and USA, 2019.
Villeneuve, Denis. *Arrival*. USA, 2016.
Winterbottom, Michael. *Code 46*. UK, 2003.

TV series

Aguilera, Pedro. *3%*. Brasil, 2016– (three seasons).
Copus, Nick. *Ice 2020*. USA, 2011.
Darabont, Frank. *The Walking Dead*. USA, 2010– (ten seasons).
Mitchell, Dominic and Jonny Campbell. *In the Flesh*. UK, 2013–15 (two seasons).

Index

Page numbers in *italics* denote figures.

Kolibri 360 GmbH
Ferdinand-Jühlke-Straße 7
99095 Erfurt, Deutschland
produktsicherheit@kolibri360.de